Robert Allen Campbell

Phallic worship

An outline of the worship of the generative organs

Robert Allen Campbell

Phallic worship
An outline of the worship of the generative organs

ISBN/EAN: 9783337283827

Printed in Europe, USA, Canada, Australia, Japan

Cover: Foto ©berggeist007 / pixelio.de

More available books at **www.hansebooks.com**

PHALLIC WORSHIP

AN OUTLINE OF THE

WORSHIP OF THE GENERATIVE ORGANS,

As Being, or as Representing, the Divine Creator, with Suggestions as to the Influence
of the Phallic Idea on Religious Creeds, Ceremonies, Customs
and Symbolism — Past and Present.

BY

ROBERT ALLEN CAMPBELL, C. E.

ILLUSTRATED WITH 200 ENGRAVINGS.

In Science, Philosophy and Religion,
The truth, the whole truth, nothing but the truth.

ST. LOUIS:
R. A. CAMPBELL & COMPANY.

PREFACE.

THE aim of this work is simply to present a popular sketch of the history, customs, and symbolism of Phallic Worship — past and present — written in plain English.

Most of the facts and illustrations given are already in print. Some of them have come down by tradition from the remote past. Many are taken from modern, and some from recent, publications. Without using quotation marks, or announcing special credits in detail, the author desires to say that he has quoted a truth, culled a fact, borrowed an illustration, and adopted an interpretation wherever found or by whomsoever before stated — and often in nearly, or even exactly, the words of the earlier writer. Those who are familiar with Higgin's Anacalypsis and his Celtic Druids, Payne Knight's Worship of Priapus and his Symbolic Language, Furlong's Rivers of Life, Inman's Ancient Faiths and his other kindred works, Lajard's Culte de Venus, Dulaure's Divinités Génératrices chez les Anciens et les Modernes, Hargrave Jenning's Rosicrucians and his Phallicism, etc., will readily recognize the sources from which much in this work has been culled.

All these works, while of the highest merit as to scholarship and reliability, are not popular; for they are redundant with masses of minutia which, while important and of essential necessity to the student making an exhaustive examination of the subject, are burdensome and confusing to the general reader. These works, too, are plentifully interlarded with multitudinous quotations, descriptions, and suggestions in foreign or dead languages — thus veiling from all but the accomplished linguist much of interest and of importance to a fair understanding of this subject.

This work is intended, then, for the honorable and intelligent general reader who desires a fairly full outline of this interesting and important department of religious, social, and political knowledge — in *English* — and without the constant veiling of socially tabooed ideas, organs, and operations in other languages.

This work is not meant for the instruction of the erudite and exhaustive student who wants a complete catalogue of facts, dates, and names. Such readers are referred to the works named above.

Nor is this book meant for the young, the ignorant, or the evil-minded; for it necessarily treats very fully, and in very plain English, upon topics and natural operations that — in this day — are denied discussion in a promiscuous assembly.

As to the importance and dignity of the theme, and hence the propriety of its treatment — which some may question; and as to its purity, which many will question — the author simply quotes Hargrave Jennings — whose learning and purity no one who knows him will question — and whose extensive and patient study of this and kindred subjects renders his opinion valuable. He says : —

" It may be boldly asserted that there is not a religion that does not spring from the sexual distinction. There is not a form, an idea, a grace, a sentiment, a felicity in art which is not owing, in one form or another, to Phallicism, and its means of indication, which, at one time, in the monuments — statutesque or architectural — covered the whole earth. All this has been ignored — averted from — carefully concealed (together with the philosophy which went with it) because it was judged indecent. As if anything seriously resting in nature, and being notoriously everything in nature and art (everything, at least, that is grand and beautiful), could be — apart from the mind making it so — indecent."

CONTENTS.

CHAPTER I.

THE FIVE GREAT SYMBOLS.

CHAPTER II.

GENERAL DIFFUSION AND MODIFIED FORMS OF PHALLIC SYMBOLS.

(9)

CHAPTER III.

PHALLIC CULTS AND CEREMONIES.

LIST OF ILLUSTRATIONS.

12 CONTENTS.

DEFINITIONS.

RELIGION is man's worship of invisible power or powers, or of an invisible being or beings — which he conceives of as like himself, but superior to himself; and which he usually denominates God — or the gods — or the divine.

Worship consists of the adoration bestowed upon this divine; of thanks for favors received and prayers for favors desired from this divine, and of obedience offered or rendered to the supposed requirements of this divine power or person — conceived of by the worshiper — as like himself, but superior to himself.

One's religion and worship will, therefore, depend upon his conception of the attributes of the divine. One's conception of the divine attributes will depend upon the unfolding and development of his conceptions of man and his attributes.

One cannot conceive of the divine with any attribute, the germ at least of which he has not recognized in man, any more than a blind man, who had never heard of light or color, could conceive of a being endowed with sensual vision.

(13)

Let the reader understand here, that this is not a statement as to anything the divine is — or may be; but simply as to man's conception of the divine.

As the ancients did not conceive of an infinite divine being, they naturally thought of a number of gods, each greater and more powerful than man, but still, like man — swayed by like motives and subject to similar limitations — each endowed with certain special powers, and with evil as well as good attributes; and always sexed — masculine or feminine. When these evil attributes were supposed to predominate in any god he was feared and avoided; and they called that being a demon.

All ancient cults — and most modern as well — recognize one among the good gods as being especially superior — the god of gods; and likewise one among the evil gods as being especially malignant — the worst of demons — a devil.

The earliest worshipers probably made or adopted some physical entities which they regarded as gods. As their ideas unfolded, these images were retained as representing the conceived of, but invisible, powers or persons which they came to think upon as divine. Then symbols were introduced to represent the images, as well as the unseen, but believed in, gods; and the gods were more fully defined. That is, images were replaced by definitions of the gods, and the statements of the divines' attributes were formulated in dogmas; and these definitions and dogmas were taught and impressed in ceremonies.

The religious world of to-day — even the Christian

world — has not outgrown these conditions. The attributes of the divine are still defined as those of a good, wise, and powerful man — only complete in aggregate and infinite in degree. God is defined as one, but there is a polytheistic personalization of his attributes as Father, Son and Spirit — each of whom have special and clearly defined characteristics, which are essentially distinct, as ruler, advocate, witness — the offended king, unyieldingly exacting justice — the merciful martyr, by works of supererogation, securing the criminal's pardon — the enlightener, making this fact and its conditions known to man. Each of these persons is in a way considered supreme in his own domain; but when, regarded as compared with each other, the Father is the head — Lord of Lords — God over all. God is defined as infinite (as if infinity could be defined), still his powers are clearly and definitely limited — not only in each of the three personalized attributes, but as to the aggregate. God is defined as masculine, and all his names — Father, Son and Spirit — are of that gender. Material images representing God are generally discarded, and by most denominations denounced; but dogmatic definitions — man-made, verbal, or intellectual images — of God are held as sacred and defended as valiantly as ever pagans protected their idols. As it is clearly illogical to define a perfectly good, wise, and powerful God as having any evil or weak attributes, these latter — which again are only those recognized in man — are recognized as aggregated in evil spirits — more wicked than men — or, as they are generally called, demons, and among whom the chief and ruler is — the Devil.

This is not written in a spirit of adverse criticism; but simply to illustrate that — the peculiarities of man's mind, which in early days multiplied gods — of comparative rank — giving them each human characteristics, good and bad — allotting to each one of them special powers and performances in the creation of man and matter — and striving, by imagery, material or verbal, to describe them and their attributes — is still man's peculiarity of mind in the foremost religion and civilization.

By phallic religion in this book is meant any cult in which the human generative organs (male or female), their use, realistic images representing them, or symbols indicating them, form an essential or important factor in the dogmas or ceremonies.

Phallic worship, in its origin and early use, was as pure in its intent and as reverent in its ceremonies, as far removed from anything then looked upon as trivial or unclean in its symbolism, as is the worship and symbolism of to-day. No people, however ignorant and savage, would deliberately allow — much less designedly introduce — any ceremony in their worship which appeared in their eyes as degrading.

The dogmas entertained by the " poor heathen " of primitive ages — which, to our enlightened minds, seem absurd, and the ceremonies by them practiced — which, in this day, would be immoral or indecent, were — to those who believed in and practiced them — as dear and necessary as are now the modern creeds and ceremonies to the more enlightened worshipers of to-day. They could not then, as they cannot now, be dislodged by denunciations.

The only way to rectify the creeds and purify the conduct and ceremonies of worship is by the enlightened and earnest teacher leading the ignorant sectarian to a higher development, so he can see the truth in a clearer and broader light; and, therefore, enabling him to interpret his old dogmas anew or to form newer and holier creeds — and hence modify and purify his worship accordingly.

Divine truth, as man sees and interprets it, is the soul of all worship — past, present, and future. As the conception enlarges and clears, the forms change, but *divine love and truth*, as man conceives of it, is the everlasting spirit of all religion. Rites which, in our eyes, are indecent, were doubtless practiced by a primitive people with the greatest purity of intent.

Indeed, it probably never occurred to the minds of these simple people that any work of nature — much less its highest and holiest activity — producing its crowning work of creation — man — could be indelicate — much less offensive or obscene.

Even the cynical and sarcastic philosopher, Voltaire, says, speaking of Priapic worship: " It is impossible to believe that depravity of manners would ever have led among any people to the establishment of religious ceremonies. On the contrary, it is probable that this custom was first introduced in times of simplicity, and the first thought was to honor the deity in the symbol of life which it has given us."

And Mrs. Child — whose intelligence, purity, and modesty needs no one's indorsement — in speaking of ancient Egyptian and Hindu religions and their symbolism, says: " The sexual emblems every where conspicuous in the sculptures of their temples would seem impure in description, but no clean and thoughtful mind could so regard them while witnessing the obvious simplicity and solemnity with which the subject is treated."

In another place she says: " Let us not smile at their mode of tracing the Infinite and Incomprehensible Cause throughout all the mysteries of nature, lest, by so doing, we cast the shadow of our own grossness on their patriarchal simplicity."

When Abraham's servant laid his hand upon the master's generative organs, in taking an oath, he was simply following the custom of the times in taking a solemn obligation. The intent was as pure, and the appeal to their recognized creator as honest, and with as little thought of indecency as in modern times we have in swearing by the uplifted hand or kissing the Bible. Jacob, just before his death, swore his son — Joseph — in the same solemn manner; and the same custom is still used among some modern Asiatic and African tribes.

The ancient matron who wore a phallic amulet, or made a votive offering to the image of an erect lingam, praying for children, was as earnest and as modest as the Jewish Sarah, Rachel, or Hannah who appealed to Jehovah; and she was as pure-minded as the modern Christian who prays to the Holy Virgin or to the Father, for Christ's sake, to give her the blessing of children. The Babylonian woman, who, in obedience to the requirements of her creed, gave herself to the embraces of the stranger who first offered her money for the temple treasury, was as earnest as any modern worshiper, and will certainly compare favorably, in purity and delicacy — to say nothing of morality — with modern wives, who would be shocked at such ornaments and procedure, and who, while enjoying all the sensual felicities of sexual congress, seek every known means to prevent conception — or to abort it even — after their preventative endeavors have failed.

Some people of our day profess religion in order to gain social standing, enlarge their acquaintance, or even increase their business; many follow Jesus for the "loaves and fishes;" and no doubt many in ancient times were pious for the sake of the sensualities; but the mass of worshipers then — as now — must be credited with pure and honest intent.

Then, as now, it was the pretenders — not those who had faith in the dogmas and god worshiped — that desecrated the rites, making them the excuse for selfish and revolting practices.

The ancients, in their worship, were not only honest in their convictions and pure in their intent, but they were careful and extended in their observations, and deliberate, as well as wonderfully discriminating in their conclusions. The foundations of essential principles which they laid and the superstructure of dogma which they erected thereon still remain in the greater part.

Only the vitality of essential truth would give such enduring life. The foundations have been deepened, broadened, and in every way improved; the superstructure has been enlarged and beautified; but the grand and eternal essentials of their cults were the germs from which have been unfolded all that we have superior to them in religion. The worship of one's creator, and the ruler of his destinies, was with them, as with us, and as it must ever be, the life of all religion.

INTRODUCTION.

THE masses of mankind, especially in religious dogmas, have always looked, as they now look, to their recognized leaders for instruction and example. These leaders have always been, as they are now, either conservative or radical. The conservative and the radical are the natural developments of two fundamentally different orders of mind, and neither class is capable of fully understanding or fairly appreciating the other class. They are opposed in purposes, plans, and methods of procedure; and are, hence, always antagonists in religion, philosophy, and politics.

Notwithstanding this continual conflict — nay, to speak correctly — in consequence of this antagonism, they are the essential and effective factors in the development of the race. They are, as it were, the centripetal and centrifugal forces in humanity. The centripetal force alone would carry the earth directly to the sun, and thus to immediate destruction by instant conflagration; while the centrifugal force alone would scatter the earth into impalpable dust, and it would be lost in the immeasurable frigidity of infinite space. So, if minds were all conservative, there would be unchanging stagnation — but no progress; and the

(21)

race would wither and die out from lack of mental
nourishment and needed exercise. If minds were all
radical there would be incessant and grinding agitation — but no stability; and the race would destroy
itself by constant and consuming friction. Yet both
these parties are essential to the existence, continuance
and betterment of the race; for just as the coördinate
operations of the centripetal and centrifugal forces in
nature causes the planets to revolve and circle in their
courses around the central sun, so it is only by the
constant activity of the conservative and radical minds,
in their opposite tendencies, and in their apparently
mutually destructive — but really coöperative — forces,
that humanity is developed in affection, intellect, and
power.

The conservatives, in religion, in their teachings,
appeal to authority, precedent, and the pronunciamen-
toes of that lamented past, when God — or the gods —
they say — walked the earth; and, standing face
to face with the wise and holy men of old, delivered
their celestial messages — which embodied all the truth
necessary, best, or possible for man to know. They
naturally formulate exact creeds, and reiterate in the
same formula of words the traditional revelations.
They insist that the time-honored ceremonies were instituted by the wise and holy fathers as a means of
pleasing God — or the gods; and thereby securing the
divine favor upon those who punctiliously and reverently observe and perform these ceremonies. They
cling tenaciously to all the old symbols. They build

monuments to the Holy Prophets of olden time — whom their predecessors in conservatism persecuted as innovators and blasphemers — but who are, now that their teachings are accepted, canonized as inspired saints. They appeal for instruction and guidance to that lamented past, from which, they say, mankind has degenerated. Their great object is, by constant reiteration of the accepted revelation, and of the established dogmas, by never flagging insistence upon the full and frequent performance and observation of all the traditional ceremonies, and by the careful and effectual suppression of all false teachings (and teachers) — as they denominate all that tends in the least degree to modify the official worship — to retard the terrible and generally inevitable retrogression from the holiness and wisdom of man's first estate; and gradually, though, of course, slowly, regain, for the faithful and obedient few, a return to paradisiacal peace. In short, they look back, they say, to the glorious sunrise of the past for enlightenment. By an unquestioning acceptance of the dogmas then formulated, by a strict obedience of the duties then enjoined, and by a full and constant observance of all the ceremonies then established, they seek to gain the special but uncertain favor of God — or the gods — they worship. They thus seek to secure, for a favorite — because obedient — few, release from the ills of this life, as well as desirable advantages in the life to come. They oppose all change of creed, lament every modification of ceremony as a degeneracy: and leave it for their children and successors to adapt themselves to the new order of things by accepting the inevitable in progress.

The radicals may, to some extent, acknowledge the truth and the authority of former revelations — for the time when it was given. They may also recognize, more or less, the time-honored traditions, as well as engage reverently in the observance of the established ceremonies. They will, however, claim that the truth was not fully revealed to the prophets of old — wise and holy though they were. At least they will claim that even if these ancient prophets were fully instructed, still we do not, from their revelations, fully understand all truth. They will assert that revelation has not entirely ceased; and will maintain that God — or the gods — will no more retire from the world as teachers than as creators and preservers. They profess to acknowledge the teachings of traditions and phenomena, but claim to look upward and onward for fuller light through intuition and new revelations. Their almost constant argument is that the asserted new revelation is in perfect harmony with the older — with all that is understood to be true and useful in the established cult. Their claim usually is, that the new light restores a lost — or brings into prominence a neglected meaning; that it unfolds an internal or spiritual interpretation — higher and fuller than the mere literal statement, or that it adds to it a new, but still harmonious, unfoldment of truth. In either case they will generally claim that there is no attempt — and no desire — to substitute a new worship in the place of the old one. On the contrary, they aim simply to develop the already accepted dogmas and practices into a clearer light and a broader usefulness.

The radicals, when wise, honest, and enthusiastic, are the real "reformers." They do not seek to substitute an entirely new authority, creed, or ceremony, but to improvingly modify — "reform" — those already accepted and in use.

True reformers, by the very constitution of their mental make-up, necessarily value more the truth than the special method of its expression; and they hold in higher estimation the spirit of the doctrines than the formal ceremonies and conventional symbols which illustrate, impress, and represent those doctrines. Their policy is, therefore, to unfold and enlarge dogmas, to re-interpret ceremonies and symbols. They seek to excise only that which the newer and clearer light shows to be false in creed, and misleading in ceremony and symbol. They aim to add only such new statements of doctrine as will express more clearly the larger truth of the new revelation. They profess to introduce only such modifications of ceremony and symbol as are absolutely necessary to more fully and more distinctly represent and impress this broader and clearer truth. The typical conservative and radical is here drawn with sufficient distinctness for the purpose in hand. It must, however, be remembered that mankind as they are — and were — range in all possible gradations of mental idiosyncrasy from the bigot — who says no change is desirable, to the fanatic — who wants everything changed — and at once.

Kings and priests — those who are in possession of inherited. vested. or permanent position. influence, or income — are, naturally. both from education and selfish interest, conservative in all things. The masses — that is. a majority of them — are not only naturally conservative. but lack the development to readily understand enlarged statements of truths. The vast majority of mankind are religious after the definition of religion, which is given elsewhere. All religion is based upon what is, according to some definitions. divine revelation. "There is no God but God; Mohammed is the prophet of God." says the follower of the faith founded upon the Koran as the only inspired and perfect revelation of Allah. the Most High. And the Mohammedan is as earnest and pious in his devotions. and as well convinced that he is a professor of the only true religion as is the Christian who accepts his Jewish Bible an l the Gospel as the only revelation of God to man : and who declares there is no God but Jehovah. and no Savior of man but Jesus. the Christ — the only begotten Son of the Father. The Brahmin. the Buddhist and the Parsee. are each equally well assured that his is the only true religion. his object of worship the only real God. and his sacred books the only truth man has received from the creator. preserver. and savior of the race.

This truth concerning the dominant cults of the present day is also true of all the minor faiths. In short, every religious teacher — from the one purest in affection and clearest in intelligence, who patiently and persistently seeks to lead his followers to worship in spirit and in truth, to the one who ignorantly and fanatically insists upon the grossest and baldest fetichism — is in our day presenting to his listeners what he believes — or assumes to believe — the truth, the whole truth, nothing but the truth, as attested by what he claims is a special divine revelation to him or his teachers.

The Christian says the Mohammedan is an ignorant bigot, accepting the teaching of a false prophet, and following the practices of a fanatical and profligate impostor. The believer in the Koran returns the compliment by calling the follower of Jesus a Christian dog, worshiping a bastard, who is admitted to be only one-third of a man. Similar insulting designations are used by each great cult for those who accept and teach a different revelation and consequently a different God.

This state of affairs in the religious world of the present time is no new, or even modern, condition of feeling and belief — of doctrine and practice. Authentic history, mythology, and the dimmest traditions of the remotest past reveal to us that man is a worshiping being; that he has always worshiped a being, or beings, which he supposed like himself, but whom he exalted as above himself in wisdom and power; that by whatsoever name or names this being, or beings, may have been known, the central idea was a superhuman wisdom and power who created the world and supervised humanity and human affairs; that the good will of this power was to be propitiated, and hence man's welfare secured by the worship and obedience of this being, while the ill-will, and hence misfortune to man, resulted from denial and disobedience

Every cult has taught that it worshiped the only true god — or gods — and that hence its followers were the favorite or chosen people — the rightful lords of creation. Every sect claimed that all others were worshiping false gods (or worshiping the true god — or gods — in an imperfect and unholy manner); that hence they were enemies of the true divine — aliens, heathens, and barbarians, who had no rights that the true believers were bound respect.

As a result of this belief, dominant and strong religious nations and sects have always persecuted the weaker "worshipers of false gods." These persecutions were graded in severity. This severity depended upon many circumstances, such as the development of philanthropy and intelligence, the comparative power of the opposing sects, and the co-operation or opposition of the civil authority. Sometimes these persecutions went as far as the extermination of the weaker "heretics," and the confiscation or even the total destruction of their property. Sometimes only the males were killed — or castrated and held as slaves — the women carried off as concubines or servants, while their property enriched the stronger worshipers of the "true god."

The faggot pile, or the headsman's axe, the confiscation of estates, and the abrogation of civil and religious rights are matters of a more recent history.

All this will illustrate why mankind are conservative from policy as well as from the natural constitution of mind.

But the mind of man is so constituted that he naturally perceives, and, therefore, must (whether he will or no, and whether or not he acknowledges the fact to himself and his fellows), recognize and accept the highest truth he is capable of comprehending whenever it is clearly presented. The uniform result of this eternal harmony between mind and truth is, that however conservative one may be in avowedly changing his creed, still the clear presentation of truth, to a mind capable of recognizing it as truth, forces its mental acceptance.

Again, man, in all stages of his development naturally loves the marvelous. To all classes mystery is fascinating. The presentation of a new interpretation, the pointing out of a new idea as embodied in an old saying, the elucidation of a transcendental meaning in a time-worn proverb — in a word the mystic unfolding of a holier purpose, a clearer enlightenment, and a greater use, in a recognized dogma or symbol, is always charming, instructive, and potential.

Different classes of conservatives may designate this unfolding as "esoteric teaching," "merely poetical," "fanciful," "impractical transcendentalism," or "nonsense."

It is, nevertheless, fascinating and effective; for even if unwarranted — nay, if it be even nonsensical and absurd — still it provokes thought, arouses the imagination, stimulates inquiry, and must result, therefore, in new and broader perception of truth.

While man cannot avoid believing the presented truth, which he recognizes as truth, still there are many reasons why he may not avow the acceptance of truth. The modest man may fear being mistaken, and honestly doubt the validity of his perceptions — especially when his acknowledged teachers refuse to accept, or denounce as false, what appears to him as true. Even if convinced he may dislike the undesirable prominence that an avowal of his yet unpopular convictions would give him. The pride of being consistent — or the vanity of being thought consistent — will prevent many an avowal. The fear of being fickle — or of being thought so — will deter many others. But, above all the fear — always well grounded — of losing favor, position, or caste among his fellows, keeps many a one from freely avowing the truth he mentally accepts.

Even some of the rulers, who were convinced by the gracious and lucid teachings of Jesus, did not openly admit the fact, because they feared the Pharisees would exclude them from the synagogue. The fear of being looked upon as unworthy in conduct on account of a change in religious connections, and especially the fear in past times — and in some places even now — of a more sanguinary and even deadly persecution, has kept — and still keeps — many a tongue from speaking a truth clear to the brain and dear to the heart. The long line of religious martyrs attest the truth of this, and those who are persecuted for " heresy " know how severe are the penalties inflicted, even now, upon all " schismatics."

The great Galilean strove assiduously to enlighten his chosen and especially intimate disciples. It is of record that he gave them esoteric instruction, which the less enlightened could not comprehend. Among his last sayings to these specially instructed followers was the assertion, " I have yet many things to say unto you, but ye cannot bear them now." But for the consolation and instruction of all his followers —(for he promised to the humblest of his followers all that he promised to his immediate disciples)— he immediately added this wonderful statement: " But the Spirit of Truth will come unto you; and when he is come he shall guide you into all truth." He identifies himself with this Spirit of Truth, and promises, that for the enlightenment and assistance of those who believe, " Lo, I am with you always, even unto the end of the world."

The professed followers (and there is no question of their integrity) of this great teacher — whom they recognize as Divine — constantly pray for the enlightening presence of this Spirit of Truth. Unquestionably religious toleration finds its highest development in the Christianity of this age and nation. Yet Christians (at least a great majority of its official teachers and prominent members — who are recognized as " pillars " in the church and strong on the faith), even in this advanced civilization, and in this age of unprecedented religious freedom, denounce with anathemas and persecute with vigor all "heretics and schismatics."

The simple truth is now — as it has always been — the great majority of official religious teachers, and their lay adherents, persecute relentlessly all "schismatics" and "heretics," punishing them with all the denunciations, pains, and penalties that their sectarian prejudices prompt, and that civil law and public opinion will allow them to inflict.

Modern religious persecution is still justified by those who practice it, just as it was in former times, by the specious, but false, plea, that the revealed will of God demands that "heathen" should be — not converted to the truth — but punished for their errors.

The crucifix, the faggot pile, and the thumb-screw cannot in this age, and in western civilization, be used to punish religious innovators; but there remains — and they are in constant use — anathemas (that is God-damnings), denunciation from the pulpit, denial of church privileges and social ostracism.

The outcome of all this is that, in nearly every community — certainly in every civilization — past and present, there were, and are, those who repeat the same creed, perform the same ceremonies, and use the same symbology, and yet give to nearly every sentence, act, and sign an almost totally different interpretation from that given by another of the same cult.

There may be in the same association — there surely is in every nation — those who, in their worship, regard the symbol merely, looking upon it as a fetich, which they fear or invoke for its intrinsic power merely.

The writer became convinced of the truth of this statement by careful and extensive investigation in the principal cities of the United States. *

On the other hand there are those who entirely lose sight — or at least cognizance — of the symbol, and looking beyond all creeds and forms, "worship in spirit and in truth" that which they think of as "the ineffable love, wisdom, and power," and which they do not assume to name — much less define.

Such worshipers are numerous in modern times, and include many who are honored for their exceptional purity, admired for their superior intelligence, and revered for their philanthropic lives. That they had representatives in the olden time might be shown by innumerable citations from ancient writings. Homer says: "Hear me, oh King, whoever thou art." Plato and Socrates are abundant in sayings which show

---

* It has come under the personal observation of the writer that one man in Wisconsin was excluded from church fellowship for cutting wood on Sunday for a sick woman. His fault was not the charitable work of providing a Sunday fire; but because he cut enough to keep the poor and bed-ridden woman warm on Monday and Tuesday. Another was excluded for teaching his Sunday-school class that he believed that a non-professor, who lived a good life, was just as likely to be saved as one who professed Christ, but lived a bad life. He has it upon undoubtedly truthful information that in Ohio, recently, a man was excluded from his church (the Dunkers) for trimming his beard round at the corners, and another for having his hair shingled — because the Bible says: "You shall not round the corners of your heads; neither shalt thou mar the corners of thy beard." A man was excluded from his church (the Amish Brotherhood) for having buttons on his coat, and a woman for wearing ear-rings.

they did not attempt to define the great first cause. Philomon writes: " Revere and worship God; seek not to know more; you need seek nothing further." Meander says: " Seek not to learn who God is; they who are anxious to know what may not be known are impious."

Every change in dogma — and consequent modification of ceremony and interpretation of symbol — is, of necessity, based upon a real or supposed larger and clearer perception of truth. It is always easier, as well as safer, for one who has this new enlightenment, to secretly read into the official creed a new meaning, and to give the established ceremonies and symbols a new interpretation, than to meet the opposition of the powers that be by any open advocacy or practice of an innovation. Many motives, commendable, permissible, and selfish, prompt—nay, almost, in many instances, force— such a course of procedure. Then, again, the order and development of mind which discovers or readily recognizes the larger truth when presented is also the order of mind which values forms as relatively of less importance than truths. It is usual, too, for those of advanced views to claim that the recognition of the larger light requires a preparation and expansion of mind which they profess to think the multitude do not possess; and this consideration will also keep many wise and prudent men from freely stating or discussing newly perceived truths.

But men, in their religious and intellectual pursuits, desire and require — as in other avocations in life —

associates of similar character and taste, as well as of
harmonious attainments, though those attainments may
be — as they naturally will be — of differing degrees.
Such men soon discover each other. They are prone
to meet together; and when confidence in each other
is established, they gladly compare views, imparting
and receiving mutual suggestion and instruction.
These meetings and discussions in past ages, when
free expression of innovating views were dangerous,
were at first, doubtless, attended only by those per-
sonally known to each other, and, of course, not in the
presence of any not known to sympathize with them.
When their numbers increased, so that the time and
place of their meetings would become noticeable, they
found it necessary, for reasons already stated, as well
as for others peculiar to their age and surroundings, to
organize a more formal association. This association
sought to increase the light they already possessed, as
well as to instruct all others who were capable of receiving their higher interpretations and purer doctrines.
The association, however, was composed of men who
were wise and prudent, as well as enthusiastic and be-
nevolent. They, therefore, sought to increase their numbers by the admission of those only who were of such
advanced intelligence as to be able to teach or appreciate (and therefore accept) the unfolding truth; of
such discretion that they would not " profane " the
sacred interpretations by stating, much less discussing,
them before those who were unable to recognize their
worth and beauty — and, therefore, " unworthy " to

receive them; and of such fidelity that they would not
betray the association, or any of its members or teachings.

The founder of Christianity selected and instructed
his disciples on principles similar to those upon which this
society was organized. He taught the multitudes by allegory and parable, as they were able to hear — that is, understand. When he was alone with his disciples he ex-
pounded all things unto them, " because," he said to
them, " unto you it is given to know the mysteries of
the kingdom of God, but unto them that are without,
all these things are done in parables."

The prime object of this association was not, as has
been plausibly maintained by some, to veil the truth from
the masses, retaining it as the means of personal gratification, and for profitable use, in the close corporation
of a select and selfish few. The grand purpose was to
develop the truth to broader dimensions and a clearer
light; to unveil it to all who could appreciate and receive it — and, therefore, be benefited by its possession; to insure that those who entered upon its study
would, so far as they were capable, continue and complete their labors; and to prevent the profanation of
the truth by its misuse. These associations gradually
developed into secret societies, composed of members
whose fitness as to intelligence, fidelity, discretion, and
courage was not only vouched for by members of the
society who knew them, but who were tested by examination and trial, and who were solemnly sworn to se-
cresy, under painful penalties for any unfaithfulness.

This was the origin of the Ancient "Mysteries;" and, in fact, of all subsequent secret societies. Whether there was only one original organization, and the others were all or mainly descended from it; or whether there were independent orders originating in different places under similar circumstances, cannot now be definitely decided. Each view is advocated by intelligent students who have given the subject patient and seemingly exhaustive study.

Alexander Wilder, whose natural bent of mind and scholarly attainments peculiarly fit him for the patient and exhaustive study he has given this matter, says:—

"It is not practicable to ascertain with certainty when or by whom the Ancient Mysteries were instituted. Their forms appear to have been as diversified as the genius of the worship that celebrated them, while the esoteric idea was so universally similar as to indicate identity of origin. In some were performed the rites of the *Bona Dea*, the Saturnalia, and Liberalia, which seem to have been perpetuated in our festivals of Christmas, the Blessed Virgin, and St. Patrick; in Greece were the Eleusinia, or rites of the Coming One, which were probably derived from the Phrygian and Chaldean rites; also, the Dionysia, which Herodotus asserts were introduced there by Melampus, a *mantis*, or prophet, who got his knowledge of them by the way of the Tyrians, in Egypt. The same great historian, treating of the Orphic and Bacchic rites, declares that they 'are in reality Egyptian and Pythagorean.' The Mysteries of Isis in Egypt and of the Cabeirian divinities in Asia and Samothrace, are probably anterior and the origin of the others. The Thesmophoria, or assemblages of the women in honor of the Great Mother, as the instituter of the social state, were celebrated in Egypt. Asia Minor, Greece, and Sicily; and we notice expressions in Exodus (xxxviii : 8), Samuel (I–ii : 22), and Ezekiel (viii : 14), which indicate that they were observed by the Israelites in Arabia and Palestine. The rites of Serapis were introduced into Egypt by Ptolemy, the Savior, and superseded the worship of Osiris; and after the Conquest of Pontus, where the Persian religion prevailed, the Mysteries of Mithras were carried thence into the countries of the West, and existed among the Gnostic sects many centuries after the general dissemination of Christianity. The Albigenses, it is supposed, were Manicheans or Mithracising Christians. The Mithraic doctrines appear to have comprised all the prominent features of the Magian or Chaldean system. The Alexandrian Platonists evidently regarded them favorably as being older than the western systems, and probably more genuine."

From the very nature of the case we can have but little direct information as to the special dogmas taught, the ceremonies practiced, or the higher interpretations of the symbols used in the secret proceedings of the "Mysteries."

THE ELEUSINIAN MYSTERIES

were the most celebrated, and are the better understood. What we can learn concerning them may, therefore, serve as a general type of all the others. Although position, influence, and wealth, no doubt, had their influence in recommending a candidate, they were certainly not all-sufficient; for Nero could not, by persuasion or threats, secure admission. Persons of all ages and both sexes were admitted.

One must have had much to recommend him before he was even thought of as a possible member. If searching inquiry concerning him resulted satisfactorily he was formally announced as a "candidate." If he was chosen, he was, under the most solemn vows of obedience, study, and secrecy inducted by a purification — including much fasting — into the Lesser Mysteries. As a concluding part of the ceremony the candidate was instructed, by the Hierophant, to look within the chest or ark which contained the mystic serpent, the phallus, the egg, and grains sacred to Demeter. The epopt then, as he was reverent or otherwise, "knew himself" by the sentiments aroused.

The real seer beheld in these emblems the symbols of divine and infinite generators — towards whose nature he aspired; the sensual and unregenerate natural man saw the representations of that which his lust hungered for. Plato and Alcibiades were aroused by very different emotions. He thus became a Neophyte — new-born, or mystic — a veiled one. He then passed a probation of from one to five years in study and purification. During this period he was subjected to various and frequent severe trials of his obedience, fidelity, courage, and discretion. When he had proven himself every way worthy as to character, and his mind was properly prepared for the reception of the higher truths, the Neophyte was conducted into the inmost secret recesses of the temple, and initiated into the Greater Mysteries, becoming a " Seer " or " Initiate." Into some of the interior mysteries, however, only a select few were ever admitted.

He was then instructed in the essential principles of religion — " the knowledge of the God of nature — the first, the supreme, the intellectual — by which men had been reclaimed from rudeness and barbarism to elegance and refinement, and been taught, not only to live in more comfort, but to die with better hopes."

This shows that the Initiates were acquainted with a higher and clearer view of the Creator, and of the present and future life, than the masses could probably comprehend.

These truths were taught, in part at least, and illustrated by " allegories — the exposition of old opinions and fables"— and by symbols. The last offering made by one initiated into the Greater Mysteries was a cock to Æsculapius.

From among the initiates some were selected who were " crowned " as an indication that they were authorized to communicate to others the sacred rites in which they had been instructed. That is, they were made, as it were, priests or teachers for those initiated — but who did not remember or understand all they had seen or heard in the ceremonies.

The Hierophant who presided was bound to a life of celibacy, and also required to devote his entire life to his sacred office. To reveal any of the secrets of the Mysteries was adjudged as the basest wickedness;

and in Athens was punished by death. Uninitiated persons found unlawfully witnessing the ceremonies were also put to death.

" The intention of all mystic ceremonies is to conjoin us with the world and with the gods." The grand consummation sought for in these initiations was, " *Friendship and interior communion with God, and the enjoyment of that felicity which arises from intimate converse with divine beings.*"

A most interesting study to one who can appreciate without prejudice that two good and intelligent men can honestly differ most radically on the meaning of a simple myth, and the ceremonies illustrating that myth, would be to carefully follow Alexander Wilder and Thomas Taylor in their essays upon Eleusinian and -Bacchic Mysteries; and then turn to the denunciation and bitter abuse of these same ideas and proceedings by celebrated and honest writers, who find in them only incarnated folly, ignorance, and worse than beastly sexual abominations.

The Initiates in their public worship professed the same creed, engaged in the same ceremonies, and used the same symbols as the masses. It is, therefore, almost certain that their private work was simply an esoteric instruction or deeper interpretation of these externals of their religion. Very gradually the permanently vital part of these secret teachings became the reformed beliefs of the masses and were incorporated into the publicly accepted dogmas. The consequence of this was the gradual re-interpretation of some ceremonies, and, little by little, the modification of such others as were supposed by their dramatic action to teach something radically inconsistent with the newer and broader recognition of truth.

As symbols have no intrinsic religious meaning, but depend entirely for their value upon the arbitrary signification bestowed upon them, they were naturally retained in their established form, while their traditional interpretations were so enlarged as to harmonize with the broader teachings of the clearer truth. The student of religious history and development finds that creeds change very gradually under the influence of increasing intelligence and varying circumstances, and he has little trouble in tracing their relationship and growth; that ceremonies, while they are modified in form to illustrate and impress the changed creed, are always a compromise between the traditional custom and the innovating dogma, generally retaining the familiar dramatic elements as well as the time-honored times, seasons, and "high days;" and that the original symbols, which represent the fundamentals in religion, remain nearly the same, the change being almost wholly in interpretation — which is the greater unfolding of the original teaching. The innovating ideas, the changed mode of thought, the new and ever-shifting conditions and circumstances of life, together with man's natural love of novelty and variety in modes of conception and expression, will evolve many new symbols and numerous modifications of those already in use; still the old and reverenced symbols remain, and in nearly the same form.

Man has, from the earliest times, recognized that every effect must have a cause. He has constantly seen phenomena which he could not, by himself, nor with the assistance of his fellow-men, either reproduce or prevent. The fact of unseen power or powers, superior to his strength and beyond his understanding, was, therefore, forced upon his attention. These unseen powers he naturally thought of as attributes of unseen beings, whose purposes were carried out with a will stronger — and often contrary to his own; whose plans were broader and more intelligent than his mind could understand; and whose operations were always superior to his comparatively puny efforts. With the first crude conception of this grand idea — which is the essential foundation of all religion, philosophy, and science — man desired to know more of these unseen and superior beings. They were recognized as at times beneficent, sending warmth, rain, food, peace, and other good gifts; and, again, as being malevolent, sending storms, pestilence, famine, war, and other disasters. Man ardently desired to know the character, purposes, plans, and powers of these superior beings, so as to court their favor, coöperation, and help, as well as to avoid their displeasure and consequent opposition.

These unseen and superior beings were thought of as personalities, who, like men, were of varying dispositions, good and bad; as of relative intelligence, some much wiser than the others; as of different powers; and as limited in locality, as well as in other respects. They were, therefore, thought of as frequently having different, and often contrary, purposes, which brought them into contention with each other. Like men, too, they were of different rank, honor, and station. They were, however, divided into two general classes — the good and the bad, those who were friendly to mankind — desiring to show him favor; and those who strove to injure, annoy, and destroy humanity.

One among them was generally considered far superior to all the others in goodness, intelligence, and power; and this supreme being was called the God, or Great God, while the others were called, simply and collectively, the gods. This supreme being, and a few of his chief associates, were also given individual names. This superior being was masculine, the creator of all that is, the father, not only of men, but of the other gods, whom he dominated. All these gods were conceived of as masculine, like the principal one. They had, however, goddesses for associates — the superioress of whom was the consort of the ruling god. These gods and goddesses were not only thought of as distinctly masculine and feminine; but they were considered as remarkable for their virility as for their other superhuman attainments. Their amours and creative endurance and activity forms an important part of all mythology. The bad gods, while inferior to the good ones, were superior to man in wisdom, strength, and virile activity; and had, also, goddesses for consorts and associates. The evil gods and goddesses, however, were destructive rather than creative; and the evil goddesses play a very inferior role in all myths. The supreme masculine creative power, principle, or person, by whatever name known, and whatever his recognized attributes, was the great object of worship and veneration; and whatever measure of reverence was shown the others, was bestowed upon them as the associates and assistants of the " Lord of Lords."

The supreme feminine creative power, principle, or person, by whatever name designated, or whatever her recognized attributes, was considered the consort or favorite associate of the masculine creator, and shared the honors bestowed upon him. This honor was in a few isolated cases, as to time and place, greater than that bestowed upon the royal god. In a greater number of instances they received equal honor. Generally, however, while they were nominally equal, the creative god was considered the wise and powerful ruler who was feared, and who, hence, received the greater share of dogmatic ceremonial and recognition; but the creatress goddess was looked upon as the tender and loving mother, and received the sincerer affection of the humble worshiper, who appealed to her as the more likely to sympathize with and assist her needy and suffering children.

Even in this day we see the same principle carried out in the purest religions. The Buddhist devotee, the pious Catholic, and the penitent Protestant, all laud the greatness, power, and wisdom of the masculine Father; but look to the immaculate Devi, the

Holy Mary, or the transcendent womanly love of Jesus, for special favors in times of unusual trial and deep tribulation.

While the above is a general outline of the supposed character and relative rank of the unseen gods, it must be borne in mind that each civilization and sect of worshipers attributed to each of the principal deities, modified qualities, purposes, and powers; and sometimes changed their rank, actually and relatively.

In India the divine fatherhood was the ineffable Brahm, or *great one*. He manifested him-herself (for he is androgynous) first as Brahma, the creator. From the latter proceeded Vishnu, the preserver, and Siva, the changer. The latter is the creator and destroyer of mankind. His destruction, however, is not annihilation, but change, hence generally, improvement. The divine motherhood — (also in Brahm) is manifested in the *mothers* or *Sactis* — Saraswati, Lakshmi, Parvati, or Devi, who are the consorts of the masculine trinity. The latter, as the wife of Siva, is the mother of mankind. This religious system is by all odds the most extensive in myth and dogma, the most finished and consistent in theology, the most elaborate and dramatic in ceremony, and the richest and most poetical in symbolism of any cult in the world. It was probably the earliest in origin, has certainly been the most presistent in continuity, and is claimed by its adherents to be — and thought by most scholars to be — the origin of all other systems. It is as Brahmanism and Buddhism to the orient what Judaism and Christianity is to the occident. There should certainly be no quarrel between these two transcendent systems, for the ethics — spiritual, moral, and philanthropic—of Siddartha and Jesus — the Buddha and the Christ — have not been improved upon. Whatever of uncleanness, dishonesty, or cruelty may be practiced by the professed followers of either of these illuminated instructors is certainly contrary to their transcendental precepts and pure examples; and whatever of purity, usefulness, and brotherly love may be developed or exhibited by regenerating men, will be only the realization of their divine teachings and philanthropic lives. They each taught a Supreme Being of infinite love, wisdom, and power, revealed the beauty of holiness, brought life and immortality to light, announced and enforced the eternal fatherhood of God and the universal brotherhood of all men. They each set the example of worshiping the Highest by giving their lives for mankind, teaching that the purest praise — most acceptable to the Divine Creator — was needed service rendered to his humblest children — the sick, the hungry, the suffering, and the outcast.

CHAPTER I.

THE FIVE GREAT SYMBOLS.

THE PILLAR, TRIAD, TRIANGLE, CROSS, AND SERPENT.

THE PILLAR.

WHY were these emblems chosen as the symbols of religious ideas? What did they originally represent? When were they first adopted? Why are they in such general use? What do they mean now? When, how, and why were the meanings of these symbols changed from their original value to their present interpretations? Why have these forms been so tenaciously retained, while their significations have so frequently and so radically been modified?

An answer to these questions will be not only interesting historically, but instructive in a more vitally important department of human knowledge — man's spiritual development. Answering these questions, even in the brief and general way which a work of this size will permit, shows that the fundamental idea of all religions is the worshiping of the Creator. Such answers will also illustrate the many and persistent oppositions which every innovation in dogma and ceremony

4 (49)

must meet, before even the fairly intelligent truth-seeker will accept them as improvements on the old creeds and forms of worship.

The early use of these symbols — dating beyond history into the dimmest traditions — their general use, among all peoples and in all times; their persistent continuance, through all the ages; their general use in our own day, when they are used by worshipers the most diverse in creed, ceremony, and life, in all stages of development — intellectual and moral — from the savage Oceanican to the cultured metropolitan, is the constant wonder of history.

PRIMITIVE MAN

was the child of Nature — the infant of the race. In the early dawnings of his twilight intelligence his thoughts were doubtless almost exclusively occupied concerning his purely physical necessities of food, shelter, and defense against his enemies — man and beast. Being the child of Nature, from whom the race, with all its improvements has developed, he, like all other children, since and now, ate his food because hunger prompted him to this pleasing satisfaction of his appetites. He put on his mantle of skin or laid it off, and walked out under the sky or sought his shelter, because his bodily comfort suggested such procedure.

The child of to-day sees its father at work "making things;" it sees its mother, or her assistants, cooking or sewing, providing food and clothing; so it can in its limited way account for the supply of its bodily wants.

So the primitive man fashioned his arrow or his garment, and hence knew how they were made. He built his hut, and captured game for his food or took it from the flocks he had cared for; so it was not a doubtful question why he was fed, clothed, and sheltered.

As the morning redness of his merely sensual thoughts were lighted up to a clearer and broader horizon by the rising sun of perception, he began to ask speculative questions as to the why and how of what he saw about him. Being a child, among the first wondering questions of a speculative nature difficult to have satisfactorily answered was, of course, the same questions which the child of to-day asks under similar circumstances.

Some morning in its experience every child's eyes are opened in wonder. There is a mystery it cannot understand. A wee bit stranger is found in the family. This baby draws its nourishment from the mother's breast, which was so recently the resting place of the now wondering and inquisitive child. The natural and anxious questions of the mystified child, so perplexing for the mother to answer with temporary satisfaction to its limited understanding, are the same questions that the primitive man asked of nature and of his neighbor — receiving only a vague, shadowy, and temporary answer. They are, too, the same questions that the scientist, the philosopher, and the theologian — even in our enlightened day of boasted research, ratiocination and revelation — ask from experience, perception, and prophecy — and from each other, without receiving any answer satisfactory to themselves, much less satisfactory to the comprehension of the inquisitive child. These questions, which every one asks wonderingly, as a child, and seriously, as a mature thinker, and which nearly every one answers glibly, without thought, and hesitatingly, as he is more intelligent, but which have never been fully answered, are these: —

Who or what is this little stranger?

Where did this little stranger come from?

How did this little thing get here?

In a word: —

" WHO MADE THE BABY ? "

These universal and ever-present questions have universal and ever-present responses, which may be formulated into universal and ever-present answers, viz.: " This little stranger is a human being. It came from God — or the gods. God — or the gods — sent it here." In short, " God — or the gods — made the baby."

Œdipus answered the riddle of the Sphinx by pronouncing the word " man ; " but he failed to solve the enigma behind the riddle, because he did not — and could not — define man. And he could not define man because he did not know himself — much less humanity.

So these formulated replies answer these questions, but they do not solve the mysteries behind these questions. They do not answer the spirit of the questions, because they do not define man or describe God. Who is he who knows man, " fearfully and wonderfully made?" and "who is he who can, by searching, find out the deep things of God, or find out the Almighty to perfection?"

All religions, past, present, and possible, must be based upon the attempt to understand and define man and God — and hence to understand and define man's relations to God and to his fellow-man. It, therefore, naturally follows that all symbology in the statement, unfoldment, and illustration, of any and every religion must have reference — directly or remotely — to the supposed character and attributes of the God — or gods — which that cult recognizes.

Man, in every stage of his development, considers himself superior to the other creatures he sees around him; he would, therefore, naturally consider his maker or creator superior to the fashioner of those creatures. Again, as man is observing before he is reflective, and scientific before he is speculative, he is prone to suppose that the immediately preceding operation is the cause of the immediately succeeding result. Primitive man readily noticed that his eyes saw, his ears heard, his tongue spoke, his hands fashioned his implements of industry and war; and he derived pleasure as well as profit from the use of these organs. His sexual organ voiced itself in his strongest passion, its appropriate activity was the source of his incomparably greatest pleasure, and produced the most wonderful and most prized result — a new human being like himself. He, therefore, naturally exalted this organ as the creator of the little stranger, who would, in his turn, become a man. Among all primitive races woman was simply a chattel, and he no more thought of giving any credit to the feminine organs, in producing the child, than he thought of considering the flint as his associate in making an arrow head. Primitive man was not yet so enlightened as to distinguish between the principle and its mode of manifestation — between the unseen force and the means of transmitting that force — between the intent that directed the instrument and the instrument itself; he, therefore, came to recognize the phallus as the creator of man.

The erect phallus was, therefore, the first object of man's adoration and worship.

Even among the earliest worshipers some of the more speculative would very soon distinguish between the phallus as a creator and the phallus as the instrument of a power which created by its use. Such men would, however, distinguish this unseen power as being masculine, and hence worship the masculine principle as the creator — still, however, using the phallus to symbolize this unseen creator.

Large men of muscular development, and aggressive natures, were the masters among their fellows. They could, and did, on this account, become possessed of more women, and hence beget more children — thus becoming of even greater renown; so stature, strength, courage, prowess, and domination became, in a measure at least, identified with virility. It was, no doubt, soon discovered that the man who had lost, or seriously injured, his phallus, was generally lacking, also, in strength, courage, and endurance. Above all, he was totally unfitted for what was then considered the great and distinctive duty and privilege of man — begetting sons and daughters. Such men were, therefore, despised and outcast. They were denied the rights of citizenship, or even the privilege of engaging in any public worship.

Phallic images, representing the organ itself, the masculine principle, or the invisible masculine creator — according to the different views and interpretations of the worshipers — were, from the earliest traditional times, made in every conceivable variation of form and size. The object presented to the eye was, from a modern stand-point of view, gross; but the idea symbolized was grand; and reverence for the creator was proved by paying abundant honor to the sign — and especially to the organ it represented. The commonest, and probably the ceremonial, or official, form, was that, however, of the erect phallus, in natural proportion, but of all sizes, from the tiny amulet — worn by pious matrons and innocent maidens as a charm, up to the imposing shaft erected over the grave of the honored hero — from which has descended the memorial columns in our modern cemeteries — and even to the gigantic phallic tower dedicated with solemn ceremonies — and the presence of which indicated a holy place — Bethel — house of God.

This phallic tower, though of course " conventionalized " in form, is still common as a church steeple, and suggests the Father of us all; while it designates a holy place, which has been, by solemn religious ceremony, dedicated as "a house of God." Our own nation — the freest in religious toleration — the wisest in philosophy — the purest in morals — the most prosperous in production that the world has yet seen, has recently symbolized its superiority by "erecting a pillar," or building a "tower," higher than the world ever before saw, to commemorate the life and virtues of its founder, and mark the world's holiest ground — the final resting place of the "Father of his Country."

THE MASCULINE TRIAD.

As men begat both sons and daughters, and as the former were much more desired than the latter, it was natural that a reason for this should be sought so that, if possible, the sex of the offspring could be controlled. As the phallus was the great object of veneration, it was, no doubt, carefully scrutinized and closely examined in all its peculiarities; but no marked difference of size, form, or condition was found that would account for the difference of begetting sons in one case, and daughters in another. It was observed, however, that men who had diminutive testicles, as a rule, lacked in virility, and that those who had none naturally — or who had lost them — were unable to become fathers. This was a revelation that the tests performed an important part in generation; and hence led to closer observation of their peculiarities. A marked and uniform difference was easily discovered. The right test is the more prominent, and hangs at a lower level than its smaller and less pronounced fellow on the left. The dimmest traditions of the remotest past, therefore, brings us the theory that the larger right testicle has the honor of giving the world its men; while the lesser one on the left has the minor distinction of being responsible for the weaker sex — a belief which is quite general at the present time in nearly every civilization.

How soon after the recognition of the phallus as creator — or as the instrument and representative of the Creator — that honor was divided with the less conspicuous, but equally necessary testicle appendages we have no means of definitely determining; certain it is, however, that the generative supremacy at first accorded to the phallus was in time divided with the tests — thus recognizing coöperation in the masuline organs of generation.

The phallus was called ASHER, which signifies to be "straight," "upright," "the erect one," "happiness," "*unus cui membrum erectum est, vel fascinum ipsum*" — "the erect virile member, charmed in the act of its proper function." ANU, probably from On, meaning "strength," "power" — especially "virile power," the male idea of creator, was the name given the right testicle, which, as the assistant in the generation of male children, was held next in rank to the phallus itself. This will readily explain why Jacob calls his son Benjamin — "son of my right side;" while the mother called him Benoni — "son of Anu," "son of my On." HOA, or HEA, — while of obscure origin, and of doubtful meaning, is clearly feminine —

and was the name applied to the third in rank — the left testicle.

The first sacred creative trinity, as recognized by the Assyrians, was, therefore, Asher, Anu, and Iloa — three distinct entities (principles or persons), each perfect in itself, each necessary to the other, working in harmony as one, towards one end — a veritable three in one — and one made up of three. In this — as in all subsequent trinities — and in fact, as in all polytheistic cults — the different organs, principles, or persons were of relative rank. One was the superior — even supreme — among the others. Their names, when spoken of or written together, were arranged in the order of their rank, beginning with the one considered as the Lord of the others — Lord of Lords. When they were spoken of as a whole, sometimes this trinity — again, like subsequent trinities — bore a name distinct from the three members, but frequently the collective unity was referred to under the name of the one recognized as highest in rank.

In comparatively later times the Jews knew and recognized this masculine triad, giving the testicles joint honor with the phallus; for their law made them sacred, so that even a profane touch was punished with death, and a man who had lost the one, or who was wounded in the other, "could not enter the congregation of the Lord." That is, a man whose creative triad was imperfect was an abomination. Even a descendant of Aaron could not be initiated as a priest if he was sexually imperfect. This rule was not confined to the benighted and licentious past, for, even in the present age of superior intelligence, one who is sexually mutilated, and, therefore, "not a man," cannot be consecrated as a priest, or promoted to a bishopric, much less, exalted to the Papal throne until an examination, both interrogative and occular — which is a part of every ceremony of ordination or promotion in the Catholic hierarchy — proves him "a man — perfect in all his members."

The same rule that religious teachers shall possess a sexual organism, perfect in form, and vital in function, is found also among some other Christian sects, and it is general in most non-Christian cults.

To this rule, however, there are some notable exceptions — ancient and modern. Some cults go to the other extreme, and require that their priests should be unmanned, either by castration or by fasting and continence. This is, of course, just as phallic as the other. Generally speaking, however, the eunuch has been, and is, looked upon with contempt — sometimes mingled with pity. This is only one of innumerable proofs we shall find that the much denounced "Phallic Worship" is by no means obsolete among the best and wisest of earth's inhabitants.

This triad was pictured plainly — sometimes moulded in plastic material, or carved as a statuette of the organs referred to in their passive condition, of natural or diminished size. A more common form, however, was a realistic representation of the organs, showing the phallus ready for active duty. These were of all sizes, from the diminutive talisman to the towering column.

No race of men, however primitive in development, or however low in the scale of intelligence, would continue long to worship the phallus, or the phallic triad, before some of the more intuitive and speculative among them would perceive that this organ was not the real creator. It would soon be recognized that, as the picture or statuette of the organs only represented the organs, so the organs themselves were only representatives of the real creator. This dawning of truth would have two marked effects upon those who perceived it; first, to develop a deeper and purer respect for the unseen power represented by the organs; and second, to introduce symbols, less realistic in form, but equally suggestive, of these organs — and, hence, of the real creative power. At what period this open and portraitive imagery began to be veiled in symbols — or how rapidly the modification was generally accepted — is unknown. The probability is that, at a comparatively early date — as is the case even now — the exoteric or realistic representations and the esoteric or veiling symbols were used contemporaneously among different classes or under different circumstances. The earliest traditions and the oldest religious relics show them both in general use.

Among the earliest modifications of phallic representations was the substitution of plain or ornamented columns, and the single upright stone— hewn or undressed — for the shafts of realistic form. These again, or rather the organ — or the creator represented by the organ — were symbolized in the single perpendicular line, Figure 2.

In the same way, under similar influences, the masculine triad came to be represented in a triune symbol of

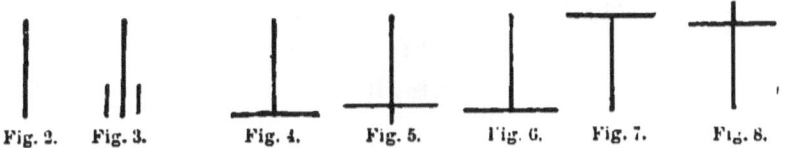

Fig. 2. Fig. 3. Fig. 4. Fig. 5. Fig. 6. Fig. 7. Fig. 8.

a single upright line with two shorter lines — one each side, as in Figure 3. This was again modified into the one upright and two shorter horizontal lines, Figure 4; which, in its next form, became the more permanent symbol of the single upright line, resting upon a horizontal line of equal length, or mortised into it and protruding through it, as in Figures 5 and 6. This was, when erected on the ground, or set up on the temple floor, not readily distinguished from the single upright shaft. It was probably to render it more distinct that the transition was made in this form to the next, by placing the horizontal bar or line at the top of the upright — as shown in Figures 7 and 8. The last four symbols seem to have been used interchangeably.

A   A verbal form of the triad which
S   esoterically contains all the doc-
H   trines of the masculine creative
E   trinity is occasionally found in an-
A N U—R—H O A   cient sculptures and is shown in its
FIGURE 9.   translated form in this diagram.

In all these representations, however, whether realistic, with all the detailed form and features of the erect virile member — or whether suggestive in the plain or ornamental column, or in the single upright stone or post — or when symbolic, in the perpendicular line, they each and all pointed to the living erected phallus. When the triad was indicated, realistically or symbolically, still the central and overshadowing feature was the erected phallus, representing creative power.

This creative power, whether regarded as the phallus, as the triad of male generative organs, as the whole man, as an unseen power, as an intelligent force, or as an intelligent and powerful, but unseen being (for we must always bear in mind that all these ideas concerning the creator have been contemporaneously held from time immemorial), was, up to the time of which we write — or rather up to the stage of development referred to — thought of and spoken of as masculine — and masculine only:

THE FEMININE SYMBOL.

Up to this stage of human development the female organs of generation, the feminine principle, the feminine creative powers, had not been regarded as factors in generation — in a word, woman had not been recognized as human; and, hence, the creator — whether principle, power, or person — was not thought of as having feminine characteristics or attributes.

Among the intelligent and intuitive men of this development there arose a new prophet, who became so enlightened that he was enabled to perceive a new and beautiful unfolding of truth. This new teacher had the seership to recognize, and the enthusiastic boldness to announce, the wonderful revelation, alike new and startling to priest and people, that the accepted and official dogmas of worship were susceptible of improvement, because a broader truth had been discovered. He announced that, while recognizing the honor and worship due the phallus, the male generative triad, the masculine principle, the masculine creator, still they were not the all, nor the all-sufficient, in generation; that important and essentially potent as is the masculine, still the yoni — woman — the feminine principle — the feminine generative power — the feminine creator — was also a factor, an essential factor, in fact an equal partner, in the generation of human beings.

Some students of ancient worship, whose patient research and eminent scholarship give their opinions great weight, are inclined to think that among the early innovators were those who not only claimed the feminine as every way equal in honor to the masculine, but carried their views to the other extreme, and exalted the feminine into the supreme place; and refused to recognize the masculine as at all worthy of coördinate worship.

The conservative priesthood and their adherents would naturally cling persistently to the old cult, denouncing the new doctrine as a blasphemous and damning heresy, and persecuting bitterly those who accepted — and especially those who taught — the worship of a strange god. The radicals would just as naturally go to the extreme of their position, and in a similarly intolerant spirit, denounce their persecutors as bigots. Each ex-

treme party would have the same form of watchword
and battle cry: "There is but one god! Our god is
god." "All honor to our god! Death to all who
worship any god but our god." These extremists, in
true sectarian spirit, waged a bitter war of words, and
carried on a more sanguinary and more destructive war-
fare of weapons. Families were divided, tribes were
broken up, nations rent asunder, in this controversy;
and not only families and tribes, but nations, were ex-
terminated in the long and savage wars which grew
out of the question of which was the true worship, that
of the phallus or the yoni — or the principles and gods
which they represented. In the meantime the great middle classes — those of the golden mean — among whom
all real reforms find their constituents — were coming
more and more to see and understand the mutual importance of the two principles, and to acknowledge both
as essential. This great middle class included all between the extremists; and their acceptance of the two
gods ranged in every possible degree of difference from
those who, while they acknowledged both gods, held
the masculine as so much superior as to consider the
conservatives practically right, to those, on the other
hand, who so exalted the feminine as to be almost in full
accord with the radicals. Still, the theory of their sub-
stantial equality in power and worthiness of worship
gradually gained ground and adherents, and finally became the dominant cult. Then, and ever since, the
worship of the creator has, in its realistic aspect or
spiritual interpretation, ranged in the same general

direction. Then, as ever since, and now (always bearing in mind that creeds, as well as ceremonies and symbols, are what we read into them, in the interpretation through them of our own feelings and thoughts), the worshiping part of mankind might be arranged under five titles, which, in the Hindu terminology, would be as follows: —

LINGACITAS,

LINGA-yonigas,

LINGAY-ONIGAS,

YONI-lingacitas,

YONIGAS.

The conservatives, who maintained the old faith, would, of course, retain the old ceremonies, as well as the old symbols of single phallus, or the masculine triad. The radicals would naturally adopt the yoni as the literal image to announce and illustrate the cardinal doctrine of the new cult. The yoni being less prominent, and hence more difficult to reproduce in full detail — the representations were of necessity more veiled. The artist, therefore, depended more upon suggestion than upon realistic reproduction to indicate the organ and all it typified. The same natural reserve which veiled a literal exposure or a picturative representation of the yoni, would also suggest other representations. The *mons veneris*, with its hirsute covering, was often substituted for the organ it concealed. This substitute is in the form of an inverted triangle; and this is the reason why the triangle was chosen to symbolize the yoni. Besides the triangle would suggest the feminine trinity — the sacred locality, the yonic orifice, and the prolific womb, and would, therefore, be an especially appropriate symbol. This triangle was usually drawn plain, as in Figure 10. It was, however, frequently rendered more literal by adding a short interior upright line, as in Figure 11.

A symbol of the yoni — and hence of the feminine principle or personality — which was common in ancient, and to some extent in modern, times — though often used with indelicate suggestiveness — was the pointed oval, Figure 12. This was sometimes softened into the ellipse, rendered angular in the lozenge, or expanded into the circle, as shown in Figures 13, 14, and 15.

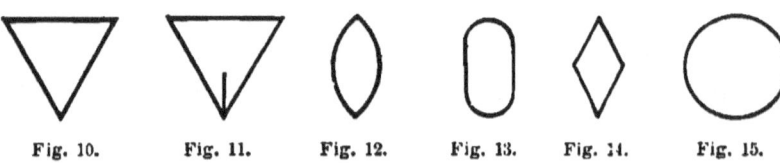

Fig. 10.   Fig. 11.   Fig. 12.   Fig. 13.   Fig. 14.   Fig. 15.

Then the woman's breast, Figure 16, with all its attributes of nourishment and beauty, was also adopted as a representative of the feminine in all its peculiarities. This was an especially acceptable and popular symbol; because it could be interpreted according to the reader's nature — either sentimentally or fleshimentally. In the swelling breast, too, the feminine triad was suggested by the two curving lines of beauty — one above, the other below — and the nipple in which they culminated. The feminine trinity was also represented by the three living colors of the pink nipple, the white field, and the intermediate band of softened tint between them, Figure 17. This representation was often symbolized by the circle as showing the out-

Fig. 16.   Fig. 17.   Fig. 18.   Fig. 19.

line of the base of the breast, Figure 18, which was again rendered more sug-gestive by placing a dot in the center to represent the nipple as in Figure 19.

The intermediate sects adopted representations and symbols which, by their forms or arrangements, or by the interpretations accorded them, indicated their pecu-liar views as to whether the masculine was superior,

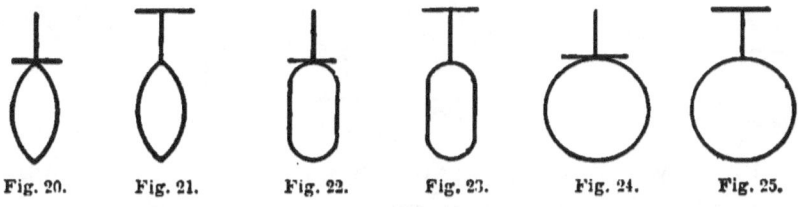

Fig. 20.   Fig. 21.   Fig. 22.   Fig. 23.   Fig. 24.   Fig. 25.

equal, or inferior, as compared with the feminine. As these doctrines were gradually adopted in different

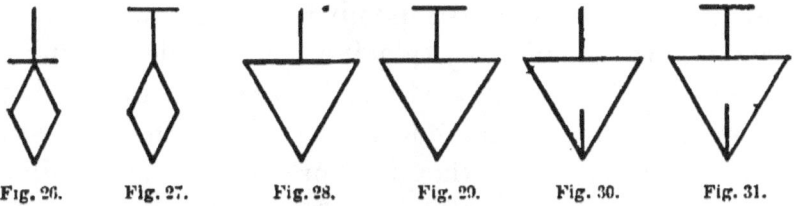

Fig. 26.   Fig. 27.   Fig. 28.   Fig. 29.   Fig. 30.   Fig. 31.

degrees, and as these symbols were used for ages contemporaneously, they were multiplied in number, and modified in form and combinations. Those who acknowledged both, but held the masculine as superior, used such symbols as Figures 20 to 31.

Those on the other hand who revered the feminine as superior to the masculine would reverse the arrange-

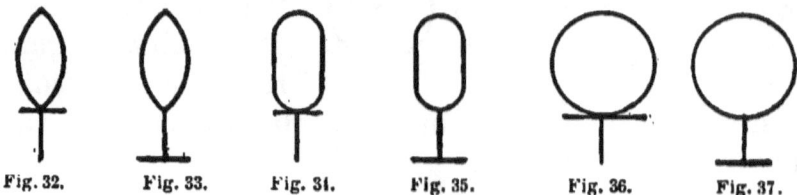

Fig. 32.   Fig. 33.   Fig. 34.   Fig. 35.   Fig. 36.   Fig. 37.

ment of these emblems and show their peculiar opinions and religious ideas in such symbols as Figures 32 to 43.

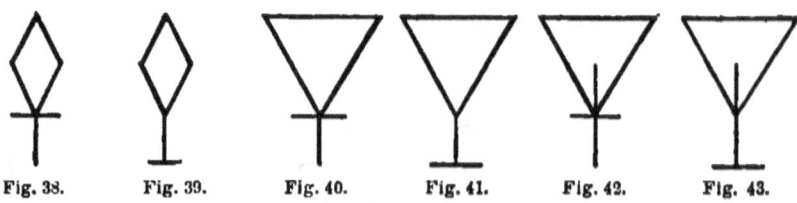

Fig. 38.   Fig. 39.   Fig. 40.   Fig. 41.   Fig. 42.   Fig. 43.

Those who contended for the equality of the feminine and masculine principles, used also the latter symbols, but interpreted them differently — saying, in substance: "We represent the masculine as a triad and the feminine as a monad or fourth member; we, therefore, represent their equality by placing the single feminine symbol over the masculine three." This class also used the symbolism of the conservatives in some cases. In the masculine triad they interpreted the upright line as the masculine and the long horizontal line as the feminine. Again, as the single upright line symbolized the masculine, this class once more adopted that sign and added their own ideas to it by placing another sim-

ilar line by its side, forming the double upright line —
Figure 48.

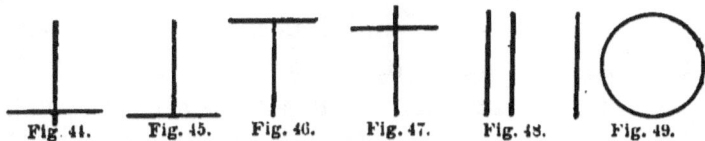

Fig. 44.   Fig. 45.   Fig. 46.   Fig. 47.   Fig. 48.   Fig. 49.

This simple symbol is unfolded by mystical interpretation in the most far-reaching application, but always
with the same generic value. Thus it means the masculine and feminine creators, Adam and Eve, Cain
and Abel, Jacob and Esau, Moses and Aaron, Jachin
and Boaz,—the two pillars at the entrance of Solomon's Temple, Peter and John, and so on, with Jehovah the Eternal Father, and Mary the Universal Virgin
Mother, as the last interpretation.

The same ideas are represented, and the same mystical interpretation unfolded by placing the upright line
and the circle together, as in Figure 49.

THE CROSS.

The race — that is the more developed part of it —
was again ready to recognize, and, therefore, to accept
a still further unfolding of the truth in regard to creation. Again, the intuitive class — who are the spiritual
eyes of mankind—furnished the seer, who, by his superior illumination, was able to perceive the new, the
needed, and the acceptable light. He recognized the
truth in the old and modified dogmas, and saw, too,
that a clearer view of these, and a larger comprehension
of their relationship to each other and to creation,

would enlarge and improve these creeds. He fully and heartily indorsed the equal importance, power, and glory of the masculine and feminine creative powers. He taught, however, that generation did not result simply from the fact that such powers are, or are equal, but from their activity in generative operation; and not from their separate and independent operation — but from their mutually reciprocal, coöperative and therefore harmoniously combined activity. This new perception of truth — illumination — revelation — call it what we may — which to us may seem a very simple and obvious truism, was to the less developed race a most wonderful and important statement; for it shed a beauteous light upon many of the obscure and, therefore, disputed elements in the already established creeds. It paved the way for sweet reconciliation between the bitterly warring sects, by showing that however the comparative power and honor of the contended-for creating principles might be regarded, that, still, each must coöperatively act with the other.

This new doctrine did not abridge the worship accorded to any recognized principle or person. It did not introduce any new object of worship. It only recognized an activity — and that a mutual activity on the part of the creators which mankind (including, of course, womankind, as well), have always recognized as a delightful occupation of their energies, and for which they were, no doubt, pleased to have a divine example and indorsement.

This new doctrine seems to have been readily recognized and generally adopted by the different sects; for while some of them contended—and in some places still contend for the superiority of one or the other of the sexual principles, all seem to hold to the necessity of their mutual, coöperative, creative work. The acceptance of this addition—not otherwise a radical change—to the dogmas, as was natural, resulted in new ceremonials made up of the old with added new features, some of which in time became—while heartily welcomed and greatly enjoyed by the worshiping participants—of a character which in this day and civilization would be denominated scandalously licentious.

The fundamental idea of the new modification of creed was the active co-operation of the seemingly opposing masculine and feminine principles and powers as

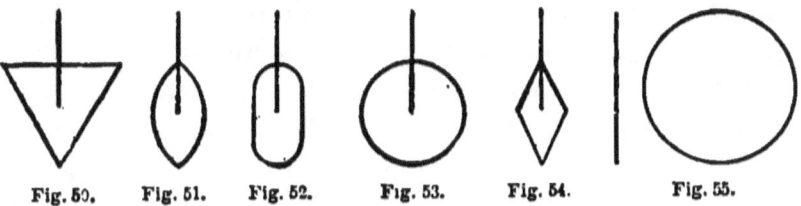

Fig. 50.   Fig. 51.   Fig. 52.   Fig. 53.   Fig. 54.   Fig. 55.

the creative cause of all that is. The imaged or pictured representation of this was naturally, of course, the masculine and feminine organs, not only in full power, ready for their special work, but actually engaged in their reciprocal and coöperative struggle to bring about the greatest of all desired results, a new creature. The symbol to veil this imagery was naturally an upright line in a triangle, pointed oval, ellipse, circle, or lozenge. Figures 50, 51, 52, 53 and 54.

The upright line and circle side by side, Figure 55, often modified into IO, and in more modern times into **IO 10**, Figures 56 and 57, are symbols almost synonymous with the cross.

Fig. 56.   Fig. 57.   They represent the union of the sexual organs — the co-operation of the masculine and feminine powers or persons. 1, the masculine, alone is simply one; 0, the feminine, alone is nothing. Their union is not $1 + 0 = 1$, but $1 + 0$ annexed $= 10$ or many. 1 is the masculine — God, alone in his majesty; 0 is the feminine — Nature, with only receptive power. 10 is God and Nature, the all-producing. 1 is the creative but invisible spirit; 0 is the existence or expression of this spirit — the visible universe; 10 is all in all, and all expressing all. The Lingacitas say 1 is all, 0 simply a servant. The Yonigas reverse this, making 0 all important, with 1 as an assistant.

But there was a sacred symbol, the combination of the upright and horizontal line, already in popular use; it was reverenced, time-honored and well understood; it was therefore policy as well as necessity to retain it. The inverted triangle, pointed oval, ellipse, circle, and lozenge could easily be replaced by the horizontal line, especially when a change of position would at once indicate the same meaning and also symbolize the new dogma. This was effected by placing the horizontal line across the middle of the upright lines, thus producing the ancient, modern, and everlasting religious symbol — the cross.

The cross, we thus see, was originally formed by the

combination of the two simplest, best-known and most transcendentally interpreted religious symbols. The upright line — the major element in the cross, still retained all its former symbolic significance as the "erect pillar." The horizontal line crossing it carried with it all the meaning of the masculine triad. Changing this line from the extremity to the middle of the upright line not only conferred

Fig. 55.

upon it, in this position, all the significance of the revered triangle, pointed oval, and circle—in a word the yoni or woman-hood — the feminine creative principle; but it did much more, for it gave both the masculine and feminine emblems and principles a living value, because it represented an active coöperative union in the work of creation.

The cross, then, when first adopted as a religious symbol meant, on the purely sensual plane, linga-in-yoni, generation by the union and coöperative activity of the sexes. It was even then, however, interpreted to signify the creation of children — on the physical plane, of course — by the orderly and designed activity of the unseen powers typed by the masculine and feminine organs. By the simple unfolding, developing, spiritualizing of this original interpretation it has come to mean regeneration — the union and coöperative activity of the masculine and feminine principles (which are variously interpreted as Divine and human — God and nature, love and wisdom, will and intellect, faith and works, etc.), to develope new creatures, who shall not only "inherit the earth" on the sensual plane, but who shall in the spiritual realm possess the heavens and fill them.

The cross has not, however, by this spiritual interpretation lost any of its interest or significance — much less had its teaching negated— on the sensual plane of man's life. Its primitive meaning and earliest interpretation is ever vital and ever present — or should be — to even the most spiritually developed. Ascetics may claim that they aim to be so busy in the work of saving the souls of themselves and others that they will have no time to engage in physical procreation; that they aim to labor so continuously and so exhaustingly in spiritual work that they will lack the power to obey their God's first command to the first parents in Paradise —"Be fruitful, multiply, fill the earth and subdue it;" that they strive to be so enamored of spiritual purity and future glory that they will have no inclination to admire the flesh or partake of its sensual felicities. There seems, to say the least, an inharmony between the teachings of God to the perfect pair, and the ascetic's ideal life of perfect man and woman now. Certain it is that if they could convert all mankind to their ideal St. Peter would have to search out some other source than Earth for a supply of heavenly inhabitants.

Another class will claim that the transcendental meaning of the cross as a symbol of regeneration in spirit should not replace — but simply supplement its interpretation as to creation on the sensual plane. They will

maintain that those who are becoming spiritually purified in affection and enlightened in intellect should even
more desire, and more persistently try, to " fill the
earth and subdue it ;" because they will give the world
a healthier, stronger, longer lived, more intellectual
and purer race of men and women—the more of whom
we have the better. They will insist that a man's desire and attempt to regenerate himself and the world
spiritually so far from negating or even interfering with
his physical duties of marriage and fatherhood, empha-
sizes not only the duties, but also augments the powers
of generation on the sensual plane ; and that however
many spiritual children he may count because of his
instrumentality in leading them from the darkness of
sin to the light of holiness, still the Lord and the world
require his best efforts to beget and bring up, on the
plane of nature, many strong sons and beautiful daughters. They will teach that the cross representing the
coöperative activities of masculine and feminine also
symbolizes that all the duties and responsibilities of
generation and regeneration equally apply to woman
as to man ; that just as virility on the sensual, and intellect on the spiritual, plane is the essence of manhood, so
fecundity and purity of affection, in their respective
domains, are the jewels of womanhood.

When the early Christian Apostles went to Egypt
and Rome — the great central homes of the new faith —
they found the cross already recognized as the supreme
religious symbol. With the same wisdom displayed by
Paul at Athens, they announced that they came — not

to tear down religious ideas or to discard the cross, but to more fully unfold the interpretations of that revered and time-reverenced symbol.   Accepting the cross and its symbolism of generation on the plane of nature — physically, they unfolded its transcendental meaning as the emblem of the divine and the human, actively coöperating to beget new creatures, that is, regenerated or divine men and women.

To write fully of the interpretations of the cross, together with its associated symbols, would be to give the religious history of the race from its primitive childhood up to its present state of comparative maturity.   To prophesy correctly its yet to be unfolded meanings would be to foretell the manner and result of man's continued growth until every son and daughter of God should attain to be perfect even as the Father-Mother in heaven is perfect.

THE SERPENT.

Probably the next new symbol, with a meaning fundamentally distinct from that of the cross, either as a whole or considered in its constituent elements, and yet representing an essential element in creation or generation, was the serpent.   This symbol of the serpent is nearly as old, and almost as nearly universal — both as to times and places — as the pillar.   No other symbol has been or is so variously interpreted.   It has meant, and is now esoterically taught to mean, nearly every transcendental truth from life to the individual on earth, and continued life of the individual and the race in the recurring generations by offspring, to the eternal life of the individual in a future and spiritual phase of existence; from simple cunning or craftiness to the broadest and clearest wisdom; and from simple sensuous light to divine illumination. The serpent has also been used to represent nearly every feeling possible to humanity, from the purely animal sexual passion to the passion of the divine man on the cross; and to symbolize every possible sensual and spiritual being, from the slimy and poisonous snake in the grass to the orthodox personal devil, who seduced our paradisical mother Eve — and who still roams the earth, seeking whom he may devour; from Lucifer — the fallen angel and prince of darkness, to Lucifer — the torch bearer of the Divine, who sheds abroad in the world all the light it has or can have; in a word from the great red dragon, the seducer of hell, the prince of error, the malignant and eternal enemy of man, to the favorite arch-angel nearest the celestial throne, the everlasting spirit of truth, the only divine instructor of man, and even the Holy Spirit — one with the Most High.

The "wise men of old," therefore, did not adopt the serpent on account of its beauty only or for ornament simply; but because they had a new and larger perception of truth and, hence, needed a new symbol to represent a new element in their philosophy. These men, being close observers, would soon notice that while the coöperative union of the sexes was necessary to, and resulted in, the bringing forth of children, which were much prized, that still this desire of procreation

was not then — as it is not now — the only, or even
generally, the main incentive to the creative act. They
doubtless recognized that if the love of the offspring —
the special desire for a child at a certain time — was the
only motive for procreative activity, that this, like
many other important duties, would often be seriously
neglected; and that, as a natural result, the earth would
be filled and subdued very slowly — if indeed it were
filled or subdued at all. They, therefore, recognized
the passion, which insured the prompt and constant
activity, resulting in populating the world, as a divine
factor in creation.   Regarding it thus, they, in accordance with their custom, sought out a representative
symbol.   They had also, no doubt, noticed that the
cobra de capella, or hooded snake of India (where the
serpent symbology probably originated) had a peculiar
power of puffing itself up — enlarging and erecting its
neck and head when aroused to excitement.   This
peculiar power, and its size, shape, position, and regu-
lar pulsations when in this condition, as well as its well-
known power of fascination — which subdues its whilom

fearful and trembling victims, were
all very suggestive. This snake,
which is the favorite form of the
earlier representations of the serpent, was, probably, for these, and,
perhaps, other reasons, chosen to
symbolize that purely selfish and
sexual passion which for the sim-

Fig. 50.

ple end of sensual gratification prompted the fleshly

union of the sexes. This significance would naturally unfold very speedily, even to the primitive race, so as to also include all those sentimental promptings which brought the sexes into harmonious and enjoyable association. Indeed the race may have been so developed as to recognize both of these interpretations from the first use of this symbol.

And the ancients were right in regarding sexual passion as divine. It is simply the divine impulse which stimulates sensual man, from purely selfish motives, and, without regard to duty or divinity, to sufferingly desire and ardently enjoy, and, therefore, to energetically and industriously engage in procreative activity. Among purely animal men — if any such there be — this passion is, therefore, instinctive — but none the less divine — in its intent and result of perpetuating the race. Among animals it is called instinct. In the vegetable world we recognize it as tendency to cellular development and multiplication. In the mineral kingdom it is known as chemical affinity. In the domain of intellect it is the spontaneous craving that seeks enjoyment in the mental activity of evolving or receiving ideas. In the realm of affection it is the anxious agitation which revels in the exciting play of the emotions. In a word this passion is, in its own domain, the special manifestation of the universal divine impulse seeking satisfaction in the reciprocal activity of creative forces ; and in man prompting him — before purity would induce or intelligence guide him — to procreative activities.

So we find that, independent of the ultimate aim of

perfecting the universe, affinity, instinct, and impulse are constantly prompting and securing the energetic coöperative activity of apparently contending, but, in reality, supplementary creative powers in the production of new creatures.

Sensual pleasure, intellectual delight, moral rapture— in a word, happiness, on every plane of man's nature—is constantly resulting from the obedience he accords to the promptings of impulse, long before he attains the moral and mental development of designedly — and with holy purpose aforethought — engaging in the same outward work.

And this impulse — whether manifested as sexual passion on the sensual plane, seeking and securing fleshly gratification; or whether it is recognized as pious fervor in the spiritual domain, longing for and ex- erting consecrated activity for regenerated emotional satisfaction — this impulse, so long as it is the sponta- neous promptings of vital strength to go forth in ener- getic activity, because that activity is self-satisfying, is what is symbolized by the serpent.

From these fundamental ideas, which the serpent has from time immemorial represented, it came to have many other significations. Its every interpretation, however, as a religious or mystic symbol has been de- veloped out of — and is the legitimate offspring of — this primitive and essential esoteric value.

CHAPTER II.

GENERAL DIFFUSION AND MODIFIED FORMS OF PHALLIC SYMBOLS.

WE cannot too fully appreciate, nor too often, in pursuing this study, remind ourselves that the use of phallic symbols, and even the use of realistic representations of the sexual organs, was, in the eyes of the worshipers using them, dignified and pure in purpose, and free from any recognized uncleanness.

THE PILLAR.

The use of the pillar in some of its varied forms was almost universal, as a religious symbol. The Teutons and Scandinavians worshiped their gods under various names, and with different attributes; but however different sects might disagree on the minor points, they all regarded the Creator as masculine, and used the phallus or its symbols as representing the Divine. The Spaniard generally worshiped a similar deity under the name of Hortanes, and used the same "staff of life" as his emblem. England, Scotland, and Ire-

6 (81)

land still bear evidence of the generality and dominancy

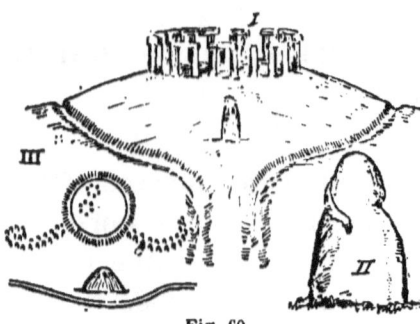

of the phallic idea in worship. To catalogue and explain the monuments and remains of this cult in the British Isles would require a ponderous volume. Stonehenge, the ground plan of which is shown in the annexed figure, has been so often written about that no description is needed.

Fig. 60.

This shows, I, the elevation as it now appears; II, an enlarged view of the " Friar's Hell;" III, the ground plan of this ancient phallic temple.

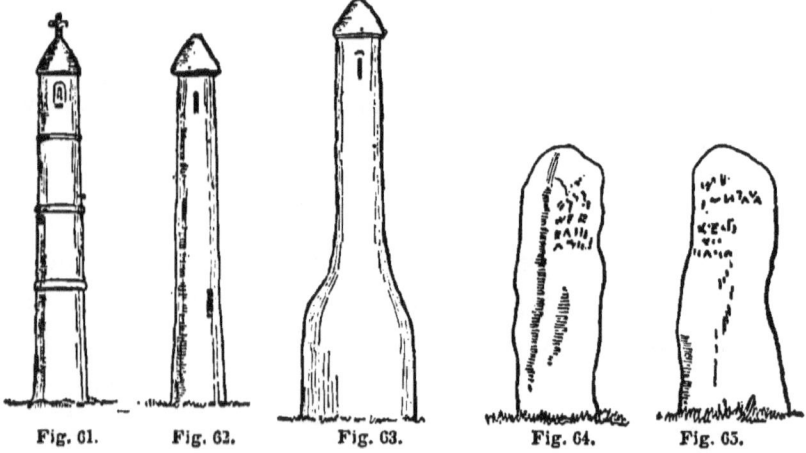

Fig. 61.   Fig. 62.   Fig. 63.   Fig. 64.   Fig. 65.

Figures 61, 62, and 63 are outlines of ancient Irish Round Towers, while two views of the celebrated Newton stone are given in Figures 64 and 65.

The pillar shown in Figure 66 is the celebrated
"R u d e S t o n e " of
Yorkshire, E n g l a nd.
The Innis Mura stone
of Ireland is shown in
Figure 67 ; and Figure
68 shows a shaft which
stands beside the ora-
tory of Gallerus, County Kerry, Ireland.

Fig. 66.   Fig. 67.   Fig. 68.

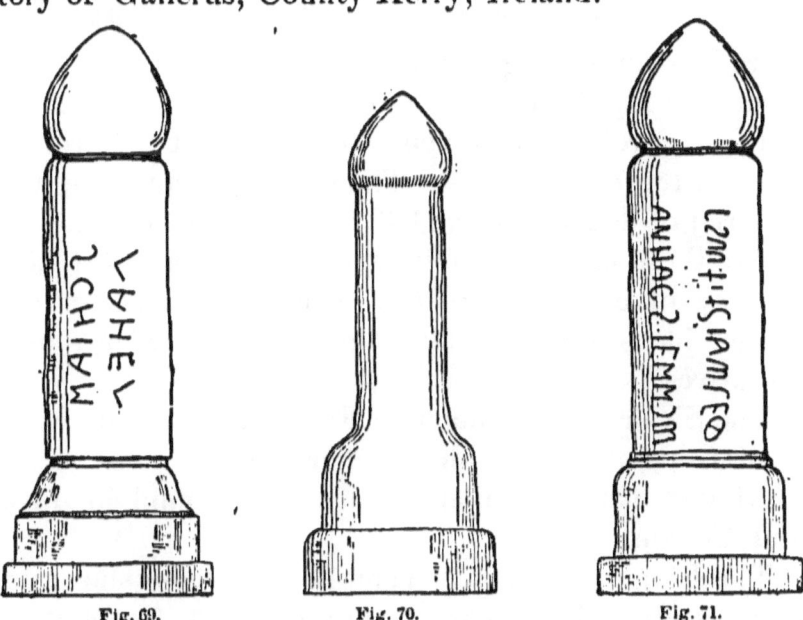

Fig. 69.   Fig. 70.   Fig. 71.

Figures 69, 70, and 71 show phallic monumental
columns found in connection with the tombs of Pompeii
and Herculaneum.

The Linga worshiped by the Parthian Magus is
shown in Figure 72. This is copied from a sculpture

found in the Baktyari Mountains. To show how wide-spread in space and time similar symbols may be found,

Fig. 72.   Fig. 73.   Fig. 74.   Fig. 75.

there is given in Figures 73 and 74 the pictures of a modern " Phallic Pillar " and " Sun Stone," as found in use as a religious emblem — or fetich, at the present time, in the Figi Islands. The shape, adornments, and material of Figures 72 and 73 are almost identical.

Are these modern emblems of the Figians any kin, by way of offspring, to the ancient symbols; or did similar ideas suggest and originate the similar representatives?

The Sivaic Shrine shown in Figure 75 needs no comment to point out its phallic character.

Almost exactly similar emblems are found in Java and Ceylon.

The Linga and Yonic Temple of India — shown in Figures 76 and 77 — are usually (at least frequently) called Buddhist Shrines.

Whether the authors are mistaken, or whether some Buddhists wander so far from the doctrines of Siddartha as to erect and use such phallic temples is not certain; but surely all idolatry and sensuality is as far from

Buddhism as it is from Christianity; for the teaching of Siddartha and Jesus are alike on the subject of idols and chastity.

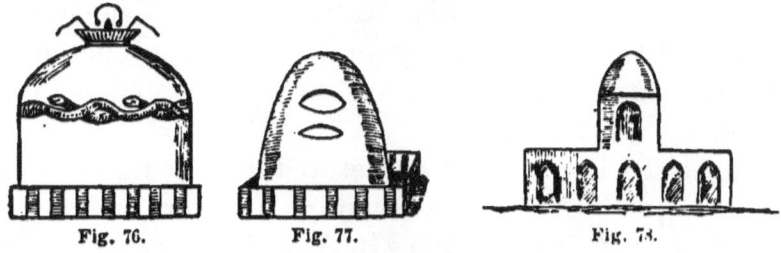

Fig. 76.   Fig. 77.   Fig. 78.

The Petrean Menhir, shown in Figure 78, is a complete combination of the masculine emblem of the "tower," with openings of a similar shape, and of the feminine "ark," or base, together with "doors" — linga in form, but yonic, from the fact of being avenues of admission.

Fig. 79.   Fig. 80.

The linga-in-yoni, shown in Figure 79, presents a very interesting example of the rude but emphatic method of a primitive people in Gothland, in expressing the recognition of the masculine and feminine principles and their coöperative union in the grand work of creation. The sacred hill at Karnak, in Egypt, the phallic character of which is obvious, is shown in Figure 80.

In a bone-cave recently excavated near Venice, and beneath ten feet of stalagmite, were found bones of animals, flint implements, a bone needle, and a linga of baked clay.

Figure 81 is a copy of a picture found at Rome when excavating the foundations of the Barbarini Palace. The mound of masonry, surmounted by the round, short pillar, is similar to those found in India, in America, and in many parts of Europe. The oval pedi-

Fig. 81.

Fig. 82.

ment and the solitary pillar have the same significance as the Caaba and hole — the upright stone and pit — revered at Mecca, long before Mahommed's time. The tree and pillar mutually interpret each other. The same idea is exhibited in modern times by two stones, Figure 82, one upright and the other with a hole in it, through which one of moderate size could pass, now found on the Island of Gozo, near Malta.

Stone phalli are common in the temples of China and Japan. Passing to the Western Hemisphere, the phallic idea is almost universal among the ancient remains of prehistoric races. In Yucatan the phallic pillar stands in front of every temple. In Panuco they adore the phallus, preserve it in their temples, and have bas-reliefs showing congress of the sexes; which is also true of Tlascala. In Honduras, the great idol is a round upright stone with two faces — the "Lord of Life," which the Indians adore; in some ceremonies they offer it the sacrifice of blood, which they draw from the prepuce. In Peru have been found ancient clay phalli, and also water jars on which were figured gods and goddesses with greatly enlarged generative organs — male and female.

Fig. 83.　　　　　Fig. 84.　　　　　Fig. 85.

In the center of the great square of the temple of the sun at Cuzco, the early European explorers found a stone column shaped like a sugar loaf, and covered with gold leaf, which was the object of special veneration on the part of the populace. Ancient phalli are found in different parts of Hayti. Figures 84 and 85 show two forms of Mexican shrines — common in the past and not infrequent at the present day.

The similarity in the outlines of these shrines or temples in Ireland, India, Petrea, Rome, and Mexico is very suggestive. In various parts of the United States there have been found excellent examples of phallic worship remains. An image found in Tennessee has an enormous phallus. Two stone phalli — one twelve, the other fifteen, inches long — were also discovered in that State. In the mounds near New Madrid, Missouri, among thousands of specimens of prehistoric pottery, there were found numerous examples of water jars exhibiting breasts and yonii of exaggerated size. These were by some supposed to be simply obscene articles; but such an idea is a great mistake — for they were found in only two kinds of localities — " worshiping places " and in burial mounds. And no race of people are so indecent and degraded as to designedly desecrate the silent city of their dead ancestors and comrades, or purposely pollute their sanctuaries.

The Antiquarian Society of Rio Janeiro, in a recently published report, state that phallic worship was common in Brazil in prehistoric times and up to a comparatively recent date; and they give illustrations of the images and symbols used in the ceremonies, and of the ornaments worn by the devotees. These are all masculine — some of them very realistic.

Phallic worship, with all the realistic emblems, is now prevalent in India, as the chapter on that country will illustrate. Mahommedan women — even in this day — reverently kiss the phallus of an idiot or a saint, recognizing them as being so holy and passionless as not to be effected by such a caress. The linga is carried in procession in Japan and in the Marianne Islands.

In Dahomey priapic figures are found in every street of their settlements. In an Egba temple Burton recently found an abundance of carvings of the masculine and feminine organs; and in the innermost sacred precinct a phallus and yoni in coition. Some natives of Africa, when traveling, carry a priapic image and pour a libation over its linga before they drink from a newly arrived at river or spring.

In some of the Pacific islands the phallic ceremonies are common. An early navigator writes of attending a native religious festival, at which a young man of fine size and perfect proportions performed the creative act with a little miss of eleven or twelve, before the assembled congregation, among whom were the leading people of rank, of both sexes, without any thought of observing otherwise than an appropriate religious duty.

The designs in figures 86 and 87 are representations of the straw pillars of the Polynesians. The s m a l l e r one, which they cover with feathers, is the more common, representing one of their gods. The "Royal God" is, however, represented by one of larger size, banded and terminating in a more realistic apex, and given a modified name showing its superiority. Straw figures are frequent in India, especially in harvest time, when they are made in most realistic sexual forms, or of human figures, exhibiting both sexes very conspicuously.

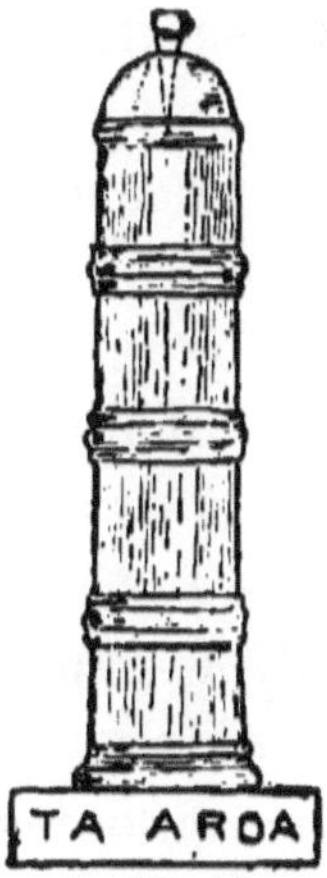

TA AROA

Fig. 86. 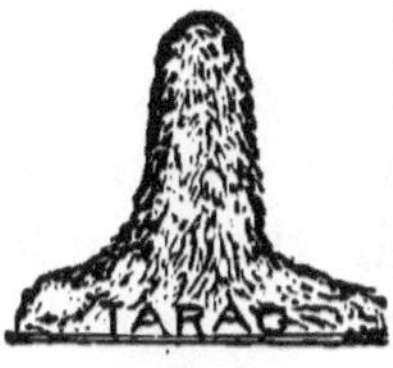   Fig. 87.

Although the Sandwich Islanders have been to some
extent Christianized; still it is well known that their old
faith frequently crops out, and there are numerous noc-
turnal assemblies, when the ancient worship of their
fathers is resumed — during which the promiscuous
and frenzied association of the sexes takes place as the
crowning part of the ceremonies. As these lapses into
phallic worship usually occur at times of threatened or actual misfortune and suffering, such as pestilence, famine,
or oppression, it would seem that the religious idea, and
not the sensual impulse, is the great motive for the appeal to their traditional gods. When the late spinster
princess, heiress apparent to the throne, died, the natives
performed their time-honored and traditional funeral
services all over the kingdom. These services were very
similar, in some respects, to the Irish wake — gather-
ings in which, during the entire night, there was feasting, drinking, and singing the praises of the deceased.
Her dominant virtue, which was universally acknowledged, and often and again extolled, was her inexhaustible virility and passion, which no man, or troop of men,
could cool; and even in her embraces with the gods she
was credited with being uniformly victorious — for she
sent them away exhausted and discomfited, because
their potency being expended, they could not accept
her invitation for repeated coitions. Similar ceremonies
are common in Africa — and in many parts of the more
civilized world. The witches sabbat of Europe, and
the Voudoo feasts of America are isolated and irregular
examples of an unregulated mixture of phallic worship,

superstition, and lewdness. Even in these latter orgies, the main purpose is not the immediate gratification of sexual appetite, but some other and more desirable favor, power, or advantage, which the leaders expect to gain by these — as they call them — religious or magical ceremonies.

THE CROSS.

There is no well defined tradition, much less any authentic history as to when or where the cross was first used as a religious emblem. Only the more prominent forms of this symbol will be noticed. The cross in the form of the letter Tau, with a circle above it, usually called the *Crux Ansata*, or emblem of life, is perhaps the most ancient. This form is very common and found in many localities remote from each other.

Fig. 88.

This form of the cross is found in most of the religious scenes depicted in the temples of Ancient Egypt. The deities — masculine and feminine — generally carry it in one hand, while in the other is the "staff of purity." It would seem that these two symbols were a constant necessity in all initiations of candidates into the mysteries. They are shown in all exaltations to the priesthood; and in the coronation of rulers. "Life and purity" were the precious gifts of the gods to kings and priests — and the treasures that the latter bestowed upon favorite assistants or neophytes. That this form of the cross, in Egypt, had other meanings than life, is

shown by its use to designate allegiance to the worship
of Isis and Osiris.   Small statues of Horus are found
also, in which he holds this symbol in his left hand, and
which there means the same as the other statues of that
god, where he holds the detached generative organs of
Typhon.   Isis is represented as holding this cross.   In
a bas-relief, from the Temple of the South, on the Isle
of Elephantine in the Nile, called the "Marriage of the
Hierophant"—that is, his initiation—the candidate
and the priestess both carry this cross in their hand.
Assyrian and Babylonian sculptures frequently exhibit
this form of the cross.   Coins found in the temple of
Serapis showed this cross prominent, and were interpreted by the early Christian fathers to mean a future
life.   Early Phœnician coins show a circular chain
of beads with this form of the cross attached — similar
in every respect to the modern rosary of the Catholic
church.   Similar rosaries are found among the Japan-
ese Buddhists, and the Lamas of Thibet.

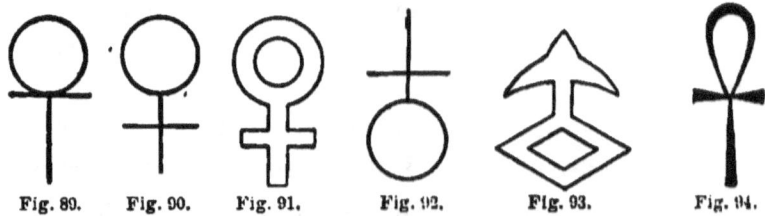

Fig. 89.   Fig. 90.   Fig. 91.   Fig. 92.   Fig. 93.   Fig. 94.

The *Crux Ansata* is also found on the ancient
Runic monuments of Denmark and Sweden — these
monuments are certainly older than the introduction of
Christianity to these countries, and were probably erected
before the Christian era.   This form of the cross is the

usual symbol of the planet Venus, as well as of the goddess of the same name. In the reversed form, as in Figure 92, it is still the coronation emblem of modern Christian countries. Figure 93 is a modification of the *Crux Ansata*. Figure 94 is copied from Pugin's Glossary of Ecclesiastical Ornaments, and is simply another modification of the Maltese cross united to the symbol of the Virgin. It is essentially the Gothic conventionalizing of the *Crux Ansata*. The Egyptians marked their sacred water jars, dedicated to Canopus, with a cross like Figure 95, and sometimes with one like

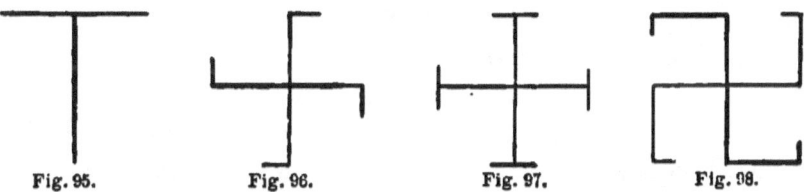

Fig. 95. Fig. 96. Fig. 97. Fig. 98.

Figure 96. The Hindus use nearly the same forms, and also one like that shown in Figure 97. The distinctive badge of the Xaca Japonicus is the cross as shown in Figure 98.

The Assyrians and Babylonians also used the cross as shown in Figures 99 and 100, to represent their "Arba-il"—"Four Great Gods"— whom they also often represent by

Fig. 99. Fig. 100.

the masculine triad in connection with the yoni.

In the cave at Elephanta, near Bombay, is a sculpturing which records the destruction of the male children in the attempt to slay Christna, and over the head of the slaughtering executioner, surrounded by supplicating mothers, is a cross.

The museum of the London University has a mummy

Fig. 101.   Fig. 102.

upon the breast of which is a cross in the form shown in Figure 101. Plato, in his Tanæus, says: "The next power to the Supreme God was decussated, or figured in the shape of a cross, on the universe." Plato drew his cross like the letter **X**.

A Pompeiian fresco exhibits a phallic cross associated with a small figure of Hermes. Ezekiel speaks of the Tau — Figure 102 — as the mark to be placed upon the foreheads of the faithful Jews, so that they might be known and spared in the massacre of the unworthy. The Tau, in ancient Hebrew, was written + or ×, and in Phœnecian thus, **T**.

The Greek cross — while usually represented as a

Fig. 103.

simple equal armed figure of two straight lines was not — and is not now, in many cases so simple.

Fig. 104. It has a much more expressive and realistic form — four masculine triads meeting in a yoni as a common center, as shown in Figure 104.

The Norsemen changed the form of the Tau into a

cross with four equal arms, and called it Thor's hammer, Figure 103.

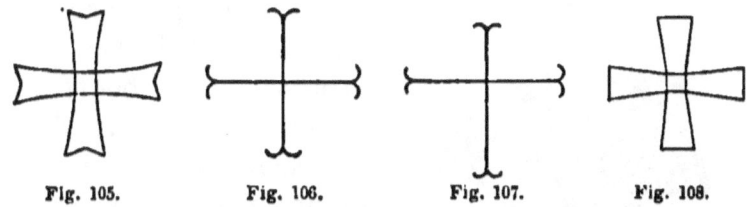

Fig. 105.   Fig. 106.   Fig. 107.   Fig. 108.

The Maltese cross, shown in Figure 105, is the same symbol. The form is more conventionalized; and, hence, more obscurely suggestive; but the character is no less phallic and triadic. Figures 106 and 107 are simply "triads" forming the Greek and Latin crosses.

The Templars' cross, Figure 108, is only a modification of the triadic Greek cross — retaining all its original significance.

Another form of the cross, similar in outline to the Greek cross, was formerly very common, but with the essential difference that the position of the organs were reversed, so that the phallus pointed outward — the other organs, however, still forming the center.

In Figure 109 is reproduced, in a conventionalized form, a copy of a golden cross, evidently worn by a person of rank, and possibly a high ecclesiastic, found at St. Agati di Goti, near Naples. In the original, the organs were figured realistically. The four arms of the cross were phalli, in erect form, pointing outwardly, the four ovals at the center were tests, and the pointed ovals at the bases of the phalli, and between them, were images in detail of the yoni, while the sacred seven was shown in the small circles form-

ing each quarter of the ornamental border; and the whole number twenty-eight represented the lunar month and the femi-nine functional month. There could, therefore, be no doubt of the phallic representation — nor of the religious symbolism, blended in mystic union; thus showing what con-

Fig. 109.

stantly recurs, the sacred and revered truths or dogmas sensuously expressed in the accentuated forms of phallic imagery or symbolism.

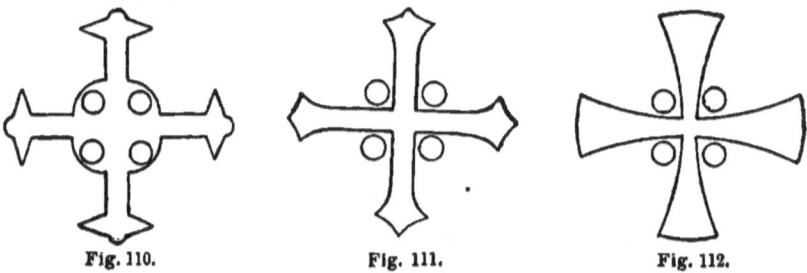

Fig. 110.   Fig. 111.   Fig. 112.

Figures 110, 111, 112, and 113 represent modifica-tions of the same ideas, and crosses in more or less conventionalized forms. In Figure 110 the major mem-ber of the triad is modified into a minor triad; while the minor members are quite realistically covered. In Figure 113 the feminine is more prominent, while in

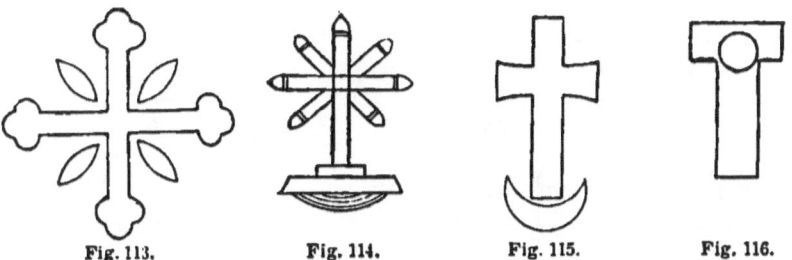

Fig. 113.   Fig. 114.   Fig. 115.   Fig. 116.

Figures 111 and 112 the conventionalizing is carried still farther.

A design often found in Greek churches, a curious combination of Christian and Mahommedan symbolism, is reproduced in Figure 115. Figure 116 reproduces the outline of a pectoral ornament worn by certain Catholic ecclesiastics in Italy about the beginning of the fourteenth century. It is simply a modification of the Greek *Crux Ansata*.

The cross was not only known and used as a religious emblem in ancient times in India, but is to-day one of the prominent symbols of the Hindu cults. The Hindus have various modifications of this symbol, the *Crux Ansata* being at once sacred and common — and esoterically interpreted in wonderful beauty. The Hindu cross (again conventionalized), shown in Figure 114, belongs to ancient days when the symbols were little veiled, while Figure 118 is the more modern form. The foundation is the same in both, but the latter has, as it were, blossomed out into a veritable "tree of life" — draped, however, in the living leaves of modern

7

delicacy, so as to veil the nakedness of its primitive ancestor.

Fig. 117.   Fig. 118.

Figure 117 presents the conventionalized form of another Indian cross, in which is shown four phalli, four yonii, and four conjunctions of the sun and moon.

The Hindus have many symbols, of the same value that the cross had originally, to indicate this active conjunction of the sexes in the work of propagating and perpetuating the race. The emblem most common

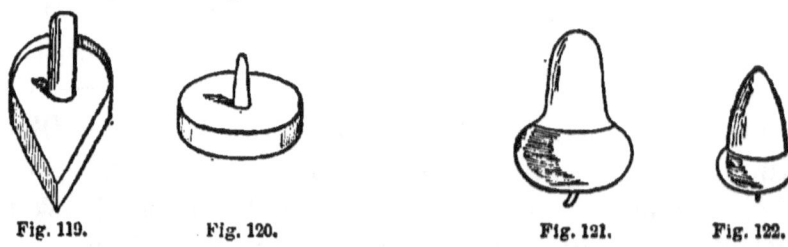

Fig. 119.   Fig. 120.   Fig. 121.   Fig. 122.

and most revered is the one presented in Figure 119. By this symbol all the others presented in Figures 120 to 141 must be interpreted.

Figure 123 is called Vishnu's Navel, and it represents more especially the creative power.

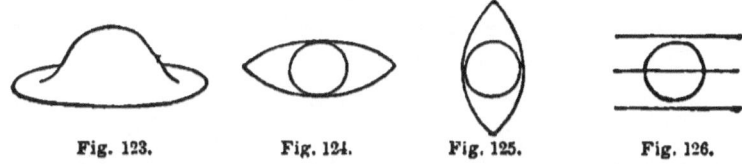

Fig. 123.　　　　Fig. 124.　　　　Fig. 125.　　　　Fig. 126.

Throughout these varying forms, which might be indefinitely increased, there is seen the constantly present

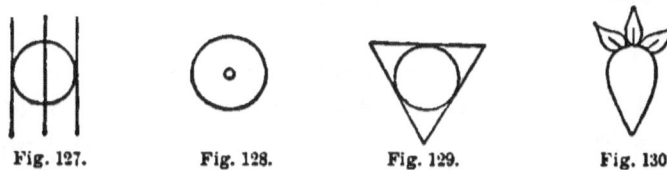

Fig. 127.　　　　Fig. 128.　　　　Fig. 129.　　　　Fig. 130.

idea of a dual and coöperative sexuality in the powers or persons represented.

This idea of the masculine — either as a unity or as

Fig. 131.　　　　Fig. 132.　　　　Fig. 133.　　　　Fig. 134.

a triad — in active union with the feminine — generally as a unity — in the work of creation is common to nearly every religious system. In Christianity it is retained in

Fig. 135.　　　　Fig. 136.　　　　Fig. 137.　　　　Fig. 138.

the masculine trinity, Father, Son, and Spirit, and the Holy Virgin. The angel told Mary " the Holy Spirit shall come upon thee, and the power of the Highest shall overshadow thee." And Joseph was similarly informed "that which is conceived in her is of the

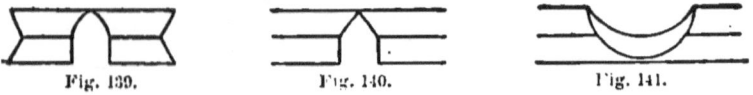

Fig. 139.   Fig. 140.   Fig. 141.

Holy Spirit." So it seems that it never occurred to Mary, Joseph, the angel, or the writers of the Gospel that a woman could become pregnant without masculine assistance.

Fig. 142.

The impregnation by which to manifest the Son is, therefore, held as true on the spiritual plane as a cause, and on the physical plane as a result. This idea is illustrated in the accompanying design, copied from a Roman Catholic "Rosary of the Blessed Virgin," licensed by the Inquisition (and, therefore, officially orthodox), and printed at Venice in 1524. The denominations, which do not, in set terms, acknowledge a feminine divine personage, as a divine creator, still teach that generation is accomplished by the use of means.

How Adam was brought into the world without a divine feminine assistant, and how Jesus was produced without a man — or why no woman was required in the first case, and why she was necessary, but a man was not, in the other — are "mysteries" not explained.

SERPENT SYMBOLS.

The symbolism of the serpent is very extensive and found in a wonderful variety of forms and combinations.

As it represents a feeling, not a thing; an emotion, not an expression; an enthusiasm, not an action; a prompter, not a performer, it is seldom found as a simple or isolated symbol. Even as a ring or bracelet it is nothing — unless worn. The meaning of the serpent must, therefore, be learned from its associated emblems; and then, also, from its form, position, and prominence, whether superior or subordinate to those that are grouped with it.

Fig. 143.

It would be most interesting to trace fully the symbolism of the serpent, but the phallic idea is the only one appropriate to the present work.

The serpent having been recognized as a sacred animal or emblem, it would, as a matter of course, be carefully studied, and all its peculiarities closely observed. As it casts its skin and thus seemingly renews its youth every year, and as it is remarkably tenacious of life, and as its bite is usually fatal — thus showing great power — it came to be recognized as an appropriate symbol of life, generative potency, and immortality.

The serpent, with his tail in his mouth, with or
 without a motto, is a very
general emblem. It originally
symbolized the passion which
prompts sexual activity, and
thus the perpetuation of the
race. It also suggests the
animal side of humanity min-

Fig. 144.

istering to, sustaining, and upbuilding the divine man-
 hood; and from these meanings it easily
came to mean immortality, future life,
and eternity. Figure 145 reproduces
a Hindu emblem of the masculine and
feminine principles united by the divine
impulse of propagation. It also signifies wisdom — intelligent, useful, pleas-

Fig. 145.

ant activity.

The linga, with two serpents twined around it, Fig-
 ure 146, is a very old and still common em-
blem. It is variously called the Caduceus of
Mercury, the Rod of Life, the Æsculapius
Rod, and the Wand of Hermes. It probably
originated in India, where it was called the

Fig. 146.   Staff of Siva, and is there interpreted to mean
the linga receiving energy and potency from the divine
influx of passion from Siva. It received this signification from the fact that the sacred serpents — the
Cobras — unite sexually in this double circular erect
form. Eastern teachers avow that it is most fortunate
for any one to see this serpentine congress, and declare

that if a cloth be thrown over them, or even waved so as to touch them, it becomes a form of Lakshmi, and therefore of the greatest procreative energy. They, therefore, preserve such a piece of cloth with the greatest care as a most potent charm in securing good fortune, and especially numerous and healthy offspring, as well as to ward off all evil influences.

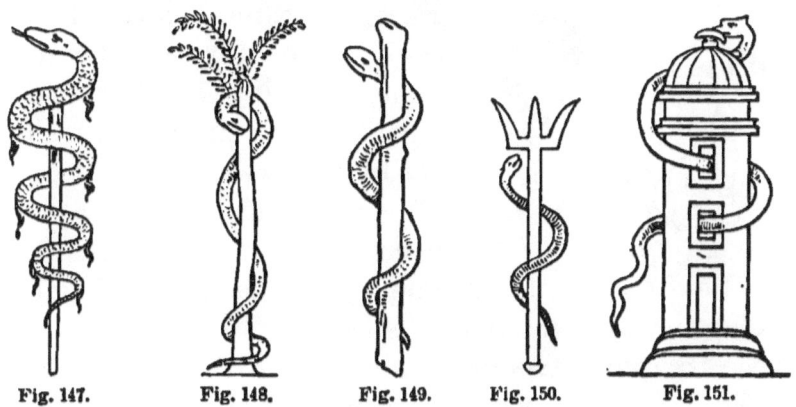

Fig. 147.   Fig. 148.   Fig. 149.   Fig. 150.   Fig. 151.

The above designs are all modifications of the "Staff of Life," energized or rendered potent by the divine impulse of vital activity. Figure 147 is a Roman Standard, symbolizing strength, vitality and enthusiasm. Figure 148 is the serpent guarding the tree of life, to keep off all imbeciles and cowards; but to co-operate with the vital, the wise and the courageous. Figure 149 is the Staff of Salvation — the emblem of Healing. In Figure 150 we recognize the trident of Jupiter, the masculine triad, the trinity of masculine creators — wise to plan, strong to execute, successful and prolific in generation. The "Fire Pillar," shown in Figure 151, is interpreted as the staff of life, permeated and guarded by the divine energy — inviting the emission that will produce new beings. Ideally, it is the Divine Creator,

Fig. 152.

sending forth the Word to enlighten by the Holy Spirit the new creatures in regeneration. The temptation, Figure 152, needs no explanation. The story of the serpent inducing the woman, and, through her the man, to eat of the tree of knowledge of good and evil, so as to become like the gods, and thus be in a situation and condition to continually develop towards the divine, is familiar to all.

The Serpent Goddess nourishing the divine impulse by which she is aroused to enthusiastic creative activity, thus increasing the number and improving the character of her children, is shown in Figure 153. The same design is also used to indicate the selfish and vampire witch who thus seeks to renew her vitality and arouse her failing passion, so as to indulge in prostituting and destructive lechery, which depletes and destroys her beguiled associates, without increasing or improving humanity. In one case the ring in which

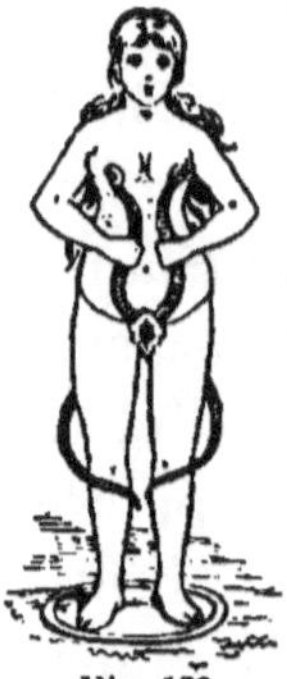

Fig. 153.

Fig. 154.

she stands is the celestial womanhood of eternal and virginal motherhood; in the other the infernal region of burning sensual desire — not only sterile, but murderous. In the first interpretation it is the door of life and the vestibule of heaven — which it is every virile man's duty and privilege to enter and occupy. In the other it is the entrance of the grave and the portal of the hells to all who therein pour their passion-poisoned seed upon a burning soil — where it is always consumed, but never germinates. Most men plant in one region or the other, and sow the seeds of humanity in soil of fertility or destruction. Momentous — nay, eternal results to the sowers and the fields — and to posterity — depend upon the choice of which door they enter, and, therefore, which region they occupy. In one case they develop purity, intelligence and power in themselves, and procreate new beings in the image of their highest ideals; and these children are born with a natural impulse toward divine perfection. In the other they are prostituting their divinest endowments, committing suicide — spiritual and sensual — and in reality murdering their possible offspring.

In the expressive design shown in Figure 154, taken from an ancient gem, the masculine creator, indicated by the sun, and the feminine associate, symbolized by the moon, are represented as brought into creative union through the impulse of divine enthusiasm, of which the serpent is the emblem. The moon being exalted, shows it to have been designed by one who worshiped the feminine as superior to the masculine — a devotee of Isis or Diva.

The following beautiful designs are also copied from ancient gems, and are each a text from which the whole system of ancient and modern ideas of creation, sensual and spiritual, might be unfolded.

Fig. 155.

Fig. 156.

In Figure 155 we have the large pillar — the divine Creator — entwined by celestial wisdom and purpose. On either side is the shell representing the female — and the tree indicating the male. The latter two are, of course, the means — the agency — the servants of the former in generating the race. In Figure 156 the tree and the two minor pillars, one on each side, represents the masculine triad. The ark below is a type of the feminine. The serpent indicates the divine impulse which secures active and creative union.

MISCELLANEOUS EMBLEMS.

The crozier is simply a modification of the original Rod of Moses or "Staff of Life," which the Jewish lawgiver adopted from his teachers when he was instructed in all the wisdom of the Egyptians. The original ecclesiastical form was that shown in Figures 157 and 158, with the double crook. Figure 159 is the more

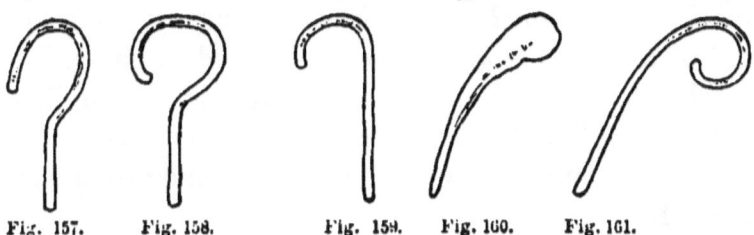

Fig. 157.   Fig. 158.   Fig. 159.   Fig. 160.   Fig. 161.

modern form. Figures 160 and 161 are varying forms with modified interpretations — expressing decorously the "Staff" or the "Instruments" which, although modernly as well as anciently worshiped — are carefully concealed in conventionalized forms.

The forked stick — Figure 162 — is another form of the same symbol, used in mystic ceremonies; and has its modern representative in the divining rod, used by the expert who "locates" water or mineral veins. It

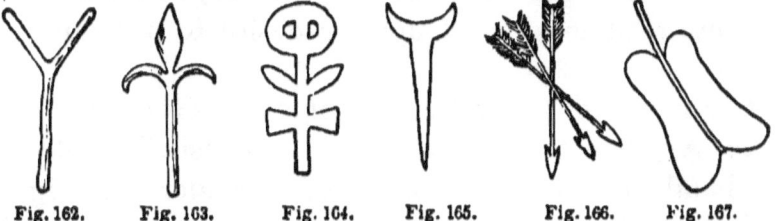

Fig. 162.   Fig. 163.   Fig. 164.   Fig. 165.   Fig. 166.   Fig. 167.

is also perpetuated in the Magician's wand. The staff with a lance head and two crooks — one small, the other large, Figure 163, is now a common emblem in India, where the women wear it as an amulet or charm to secure good luck — especially to insure children and

ward off evil spirits.   They interpret it to signify the
masculine triad or phallus — according to their ideas of
the masculine principle.   Figures 164 and 165, copied
from Lajard's *Sur le Culte de Venus*, are symbols in the
hand of a large female figure sculptured in bas-relief on
a rock at Yazili Kaia.   Figure 164 is a complicated
symbol of "The Great Four," while 165 is the masculine staff surmounted by the crescent moon — the symbol of Isis.

The arrow — Figure 166 — the emblem of Cupid, or
as he may appropriately be called, St. Desire, is synonymous with the "pillar."   The bow — relaxed or strung
and taught — is a symbol of virility spent or in abundant plenitude.   Hence, the store of the arrows in the
quiver shows a reserved stock of virility, capable of repeated encounters.   We thus see the meaning of the
composition in which appears the spent arrow, or virility
manifested — the quiver or reserve force awaiting
opportunity — the bent bow with taught string, immediate readiness, and Cupid or desire to ultimate this
force.

Fortune — or Saint Luck — or Saint Good Fortune,
is always depicted as a woman.   She usually holds in
her hand the steering "oar" or "rudder" — Figure
167 — which she offers to him who has the courage to
accept it, with the strength and skill to use it.   Such a
rudder and such a helmsman will insure speedy, oft recurring and delightful voyages, with abundant and
increasingly prosperous results.   When it is borne in mind
that the "oar" Fortuna offers is an emblem of the

"staff of life" we can easily see the beautiful appropriateness of her motto, "Fortune favors the bold."

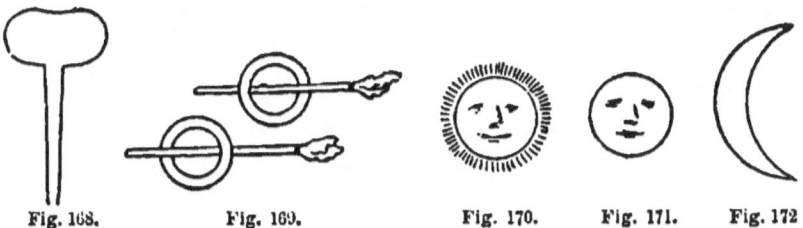

Fig. 168. Fig. 169. Fig. 170. Fig. 171. Fig. 172

Figure 168 gives a symbol less common; but the interpretation is quite as forcible. It is the "hammer" which strikes the "anvil" and forges out children. And this symbol is associated with the motto, "Every one is either hammer to strike or anvil to bear."

The "staff in the ring," shown in figure 169, is found on coins in connection with the bull — they represent the "Great Four," and are used like the above as charms.

While the conventionalized forms to symbolize the tree of life, and the masculine triad are so numerous and varied, still there are many ceremonies and occasions when the realistic forms are required and used; and wherever the real meaning of the emblem is recognized, the more realistic representations are generally supposed to be the more potent.

The sun, being credited with the active and fructifying powers of nature, was among the ancients regarded as the father, masculine principle, or God paternal, of all that is. The moon and the earth, being looked upon as receptive only — were in like manner denominated mother—feminine creative principle, or goddess maternal of all that is created. To the sun, therefore, was attributed all manly and virile attributes. David, in his psalms, shows this idea as common to the Jews, for he speaks of the sun as " a bridegroom coming out of his chamber," *i.e.*, as a virile man replete with generative vigor. The vernal equinox was celebrated as the marriage of the sun and earth. The swelling bud and opening flowers typed the promise of fruit, as the result of their consummating this union ; and the clustering grape, luscious fruits and sustaining grains were welcomed as the offspring of this celestial-natural union of the masculine and feminine creative principles.

As the sun and moon and face of nature remain the same from year to year —with apparently ever renewed life and vigor — remaining as it were in the prime of life, fresh and unchanged by age, and unweakened by use, the ancients came to think of the moon as the ever continuing virgin wife of the sun-god — and the everlasting virgin mother of all inferior deities and beings. The ancient month was measured by the interval between a new moon and the next new moon. This interval of time also marks the recurrence of the functional peculiarity of women, which ceases as soon as pregnancy occurs. The lunar crescent — new moon — probably from this cause among others — became a symbol of virginity. This is one of the most common and widely diffused emblems, and is found in most cults, ancient and modern, adorning the brow, or in some other way designating the feminine, maternal, and virginal creatress. The crescent was worn among some ancients, and is now worn in Italy as an amulet especially appropriate to virgin and pregnant women.

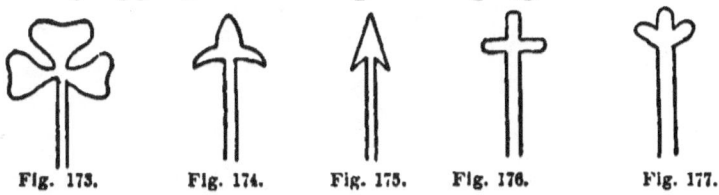

Fig. 173. Fig. 174. Fig. 175. Fig. 176. Fig. 177.

Figures 173 to 177 are each symbols of the masculine triad, and are common to most phallic cults. Figures 178 to 182 are emblems of the same idea peculiar to the Hindu religions.

Fig. 178. Fig. 179. Fig. 180. Fig. 181. Fig. 182.

The masculine creative triad is also represented by the right hand in the position shown in Figure 183. This is an emblem of great antiquity; and is found on many of the most ancient Hindu, Assyrian, and Grecian sculptures. It is the symbol of divine light, truth, authority, and mystery, by which initiates in ancient wisdom knew each other. This form of hand is placed Fig. 183. upon the head of the staff of justice in France; and is often found on the staff or wand of authority in coronations and other important exaltations. It is common in early Christian art, and is the form in which the Pope raises his hand when he blesses the faithful.

The symbol of the hands, as shown in Figure 184, 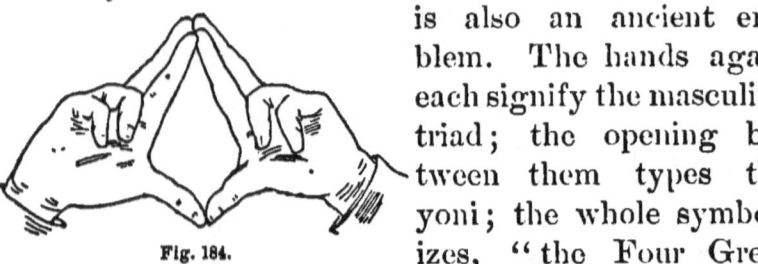 is also an ancient emblem. The hands again each signify the masculine triad; the opening between them types the yoni; the whole symbolizes, "the Four Great Gods," from whom all beings emanate. This is the form in which the Jewish rabbi raise their hands when pronouncing benediction.

Fig. 184.

In many ancient countries — and the same is true of some modern peoples — the seeing of the living yoni — especially that of a maiden — was considered the certain harbinger of good fortune.

Ceres wandered over the earth, seemingly disconsolate beyond cure. Baubo, after exhausting all other means of cheering the goddess, finally retired, shaved the hair from her *mons veneris,* and returned to the celestial presence. She then sat down before Ceres with her legs wide apart and her skirts drawn up so as to exhibit her now youthful-like yoni. This sight so attracted and pleased the disconsolate goddess that she immediately smiled with hope, partook of refreshments and renewed her gladness of heart.

This myth is interpreted to mean that philosophy or ideality alone will not produce happiness; but that the thoughts and the activities of life must engage in the sensual, as well as the ideal, in order to secure the sweetest and best results in this life as well as in the future. Without the participation of the intellect in the corporeal operations of nature, only brute life would be generated and perpetuated; and without the corporeal activities of generative energies, purity and truth would have no means of expression or of increased development in humanity.

The eye, or yoni, was placed over the portals of temples and tombs in Egypt, Sicily, and other countries, and was everywhere the emblem of life, health, and good fortune.

In Ireland, until recently, several churches bore over their main entrance the rude, but elaborate, sculpture of a woman pointing to the realistic, but exaggerated, representation of her yoni. A similar woman was sculptured on the side of the church entrance at Servatos, in Spain, while an equally phallic man was exhibited on the other side. In some other cases the key-stone over the portal bore the realistic yoni only. Similar representations were found in Mexico and Peru. It was a common practice among the North African Arabs to place over their door the genital parts of a cow, mare, or female camel — representing to them the human yoni — as a talisman to avert evil influences. There is among all peoples more delicacy about exhibiting the yoni and its realistic representations than is observed in regard to the phallus; hence, there has always been the custom of using veiled and suggestive emblems for the female organs.

8

The most popular modern representative of this yonic charm above the door is the plain horse-shoe, so common, and by many considered so potent for securing good luck. It is often associated with the cross, and frequently with the arrow, as shown in Figure 185, which is a very mystic Templar and magic emblem of the Middle Ages.

Fig. 185.

The pointed oval, or as it is called, the *Vesica Picis*, is sacred in the church, ancient and modern. It is often the frame — or rather the "door of life" —

Fig. 186.

Fig. 187.

in which appears the Celestial Mother. Figure 186 is an Indian representation of the "Gate of Heaven" — copied from a sculpture of an ancient Dagopa in the Junnar Cave, Bombay Presidency. The same idea is also represented in a modified form in the monastery at Gopach, in the valley of Nepaul, as shown in Figure 187. It is possible — but not probable — that the symbol has a horse-shoe for a frame, for the Orientals are very realistic in their illustrations. The worship of the feminine is, however, clearly shown in both designs. The sacredness and holiness of the yoni is clearly announced by making it " holy ground " by the presence therein of a deity to be adored.

In the same " Rosary of the Blessed Virgin," referred to on page 100, there is a representation of " T h e Eternal and Holy Virgin," in this almost realistic " door of life," which is reproduced in Figure 188. Figure 189 is a copy of the medal worn by the pilgrims to the shrine of the Virgin of Amadou in F r a n c e. It is commonly spoken of by those who wear it as the Mother and Child in the " door of life." Figure

Fig. 188.

190, copied from Lajard, represents Harpocrates seated on a Lotus, admiring the lozenge, as representing the Divine Mother. Such homage as is here depicted is even now paid by some sects in India, Palestine, and Africa to the living organ. The devotee on bended knees, and in silent prayer, offers to the uncovered yoni a part of the food given him by the woman, before he tastes it, which she accepts and eats, as evidence of its purity from poison. This exhibition and adoration

of the yoni is simply their method of vowing mutual
friendship ; and is similar in meaning to swearing by the
grasping of the phallus, and — like our uplifted hand

Fig. 189.   Fig. 190.   Fig. 191.   Fig. 192.   Fig. 193.

when taking an oath — is an appeal to the divine creator
as a witness of truth and good will.

The shell or *Concha Veneris*, Figures 191 and 192,
is a very common symbol of the yoni, and, hence, of
all it represents.   This is an ancient and modern symbol, often worn as an amulet.   It is common in Italy,
and is there the especial badge of pilgrims to some
shrines.

The cornucopia, Figure 193, is a similar symbol.   It
contains libations which are poured upon the phallus,
but more especially upon the prolific womb.   It hence
signifies abundant felicity, plenty, and good fortune.

The feminine hand, shown in Figure 194, is of similar

import as the shell, cornucopia and *vesica
picis* — that is, the making of this sign is
interpreted to mean that all the felicity and
blessings represented by these emblems are
wished by the signaler to fall upon and
follow the one to whom the hand thus

Fig. 194.   formed is shown.

The eye is a well known and very common symbol of Devi, and plays a very conspicuous part in many ceremonies having a phallic origin or intent. In India it is drawn plain as in Figure 195; but Ashtoreth, or Astarte, is often represented by an eye drawn in rough outline, as presented in Figure 196, and is then interpreted to mean the door of life— feminine fecundity — the Mother Creator. There is no physiological reason why the eye is any more appropriate to symbolize a goddess than a god — for sight is equally an endowment of both sexes. The eye, as drawn horizontally, is simply the *vesica picis* in a changed position from its natural perpendicularity; and the pupil represents the masculine emblem in its union therewith — that is the androgynous character of the Creator. The Indian myth explains how and why this symbol was adopted, and also explains the meaning of the spotted robe of divine personages, as well as the spots on sacred or symbolic animals. The story relates that Indra, like David, became enamored of a beautiful woman whom he accidentally saw, but who was the wife of another man. This woman's husband had, by his piety and austerity, attained to almost divine power. He forgave his erring wife (a really divine thing to do), but he punished the adulterous god of the sky by covering him with a multitude of pictures of the yoni. This was a terrible mortification to Indra; but, by the

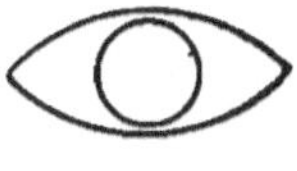

Fig. 195.

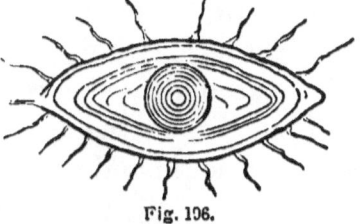

Fig. 196.

intercession of the other gods, the wronged husband was induced to change the yonii on the culprit's body into eyes. These, however, were to be so arranged in threes or fours as to preserve their phallic meaning.

The eye — the all-seeing eye — is a favorite modern symbol, especially with secret societies. It may have had its origin as above suggested; but, independent of this myth, it has a good foundation as the symbol of the Mother Creator, or as the feminine side or attributes of a masculine or androgynous creator. The ancients — and many moderns as well, considered reason — in the sense of logic and calculation — as a peculiarity of the masculine mind, while they looked upon perception and intuition as especially feminine attributes. The eye, as the organ of sight, would, therefore, naturally represent intuition, and hence the Celestial Mother.

THE COCK has from time immemorial been the symbol of masculinity. The doctrine and interpretation seems to be that the cock announces the rising sun — the god of day. For its size this bird is remarkably strong, courageous, and enduring, and he seems to have unlimited virile powers among the hens.

Minerva — also called Pallas — is often shown with a cock sitting on her helmet; and her crest denotes her admiration for this salacious bird. The sacrifice of a cock was a solemn ceremony of the highest order in Greece. The Celts also practiced the same ceremony. The sacrifice is common now in many parts of Asia, where the priests select at will — for no refusal is anticipated — the finest bird in the village. They carry it to the top of the hill and there, upon the summit, offer to the divine the sacred fowl — spattering his blood over their Tsur-oo-Salem — "Rock of Ages." Payne Knight reproduces a design in which the body of a man has for its head the body of a cock, of which the beak is a linga — the pendant wattles being the other two members of the masculine triad, and these, with the comb, suggesting very plainly the capilary adornment. The inscription reads " *Soter Kosmou* — Savior of the World," a term applied to all deities, but more especially to those charged with creative functions.

The weather cock — or its substitute, the arrow, which has the same meaning — is the modern survival of the ancient emblem. Whether on the pole, barn, or church spire — in which last place it is a peculiarly appropriate adornment — it stands forth in vital and defiant dignity, with its head meeting and dividing the wind, which is the natural emblem of the active creative feminine.

The Chinese represent the sun by a cock in a circle, and a modern Parsee will on no account kill one. The cock is a common symbol on Greek monuments.

THE TREE. — The pine tree, by its height, straightness, and evergreen foliage, was recognized as especially appropriate to represent the ideal phallus. From this it was easy to adopt the pine cone, as a masculine emblem symbolizing especially the testes, and thus energy and impregnating potency. Thus, it is easily seen why the wand of Bacchus — the thrysus — terminates in a pine cone.

The palm tree, for similar reasons, in the countries

where it was the " great tree," was used for the same purpose, and so palm branches have been used as they are now, and, in their absence, pine or other evergreens, as emblems of life, peace, and happiness.

Even within the present century the women of France, on Palm Sunday, carried in procession, at the end of their palm branches, phalli made of bread, which they called " la pine "— the French euphonism of the phallus — whence it was called *the Feast of the " Pines."* These " pines," having been blest by the priest, were kept for the year as an amulet.

The palm tree, when used as a sacred emblem, was usually conventionalized as having seven branches.

Fig. 197.

The first Jewish coinage, under the Maccabees, shown by the shekel given in Figure 197, at once tells of the palm as being a sacred tree, and also that seven branches, as spoken of in Exodus and Revelation, was likewise a revered national emblem.

For similar reasons the oak, in the countries where it is the tree dominant in size, has been dedicated to similar purposes with like interpretation.

In India the Banyan is, for like reasons, the sacred tree.

CHAPTER III.

PHALLIC CULTS AND CEREMONIES.

PHALLISM IN INDIA.

IT is questioned whether the writers of the Vedas were acquainted with — or, at least, whether they recognized or practiced — any form of worship in which the generative organs or their symbolic representations were used in any sensual way.

LINGA WORSHIP,

however, is spoken of freely in the Puranas, and one of them is called the Linga Purana. The authority for, and the origin of, Linga worship, as well as the prominence and prevalence of its imagery and symbolism, is accounted for in a myth with the following outline : —

"A powerful company of wicked conspirators, whose hypocrisy Siva had exposed, sent a consuming fire to destroy the genital organs of the latter. Siva was so indignant at this attempt to unsex him that he threatened to destroy the human race. Vishnu implored him to suspend his wrath. Siva relented in his purpose of extermination; but ordained that in his temples those parts which his enemies had attempted to destroy should forever be worshiped."

The Eastern devotees not only obey this ordinance, but go farther and model the architecture of their temples after the phallus, as the divinely formed and indispensible medium ordained by God himself for human propagation.   Lucian speaks of such a phallic temple of great height as existing in Syria.   The primitive linga is said to have been a radiant pillar in which Mahesa ("whose form is radiant as a mountain of silver, lovely as the crescent of the new moon, resplendent with jewels,") dwelt, and on which was visible the sacred word OM.   How suggestive this is of the pillar of fire in which Jehovah went before the Israelites.

The linga is always found in the Hindu temple.   It is generally in the sanctum, or holy of holies, and is made of granite, or other stone, ivory or precious wood. On certain occasions it is garlanded with flowers ; sometimes above it is a brilliant golden or silver star.   On great occasions it is honored by a light from a seven branched lamp.   The same emblem, smaller in size, carved in gold, silver, ivory, crystal or sacred wood, is worn about the neck, in the turban, or in the bosom, as a charm, or amulet — and as a declaration of faith.   The Hindus use it in prayer as the pious Catholic uses the image or symbol of his patron saint.   It is also often buried, by request, with the body of its former owner. Worshipers of Siva also mark his symbol — an upright line — on their foreheads ; while the followers of Vishnu use a horizontal line with three short perpendicular lines.

There is much misapprehension in regard to Siva,

who is often spoken of as the god of destruction. This is a misleading name. He is not the creator of original matter, but the diety who makes new forms — or new beings — by the process of changing the old. He is in fact the god of evolution. Hindus look upon change as the cause of suffering, and, hence, they long for Nirvana, which is "changelessness." Still even Nirvana is attained by change. So Siva might be called, in Western phraseology, the god who develops by discipline.

Siva, the diety presiding over generation, is the god especially worshiped under the form of the linga; but as in other cults of similar nature he is symbolized by, or ideally seen in, all pillars, obelisks, pyramids, high trees, limbless trunks — especially palm trees, poles, upright lines, high places, and in the triangle with the apex upward. The linga pillars are of all sizes. Some of them are gigantic. They are usually red, but frequently of other colors; some being black, and the one in the golden temple at Benares is pure white. The principal seats of linga worship at the present time are in Northeastern and Southern India. As these are localities little under Brahman influences, it tends to show that this form of worship preceded the Brahman religion.

The temples of Siva worship are in many parts of Hindustan — especially along the banks of the Ganges — more numerous than those of any other religion. Benares, however, is the great center of this form of worship. The principal diety there is Visweswara, "the Lord of All." His symbol is a linga; and most of the objects of pilgrimage are kindred stone symbols.

These temples are square buildings with round roofs tapering to a point. In Bengal each one consists of a single small square chamber surmounted by a pyramidal center. The linga occupies the center, and the offerings are made on the threshold.

Strangers are not, of course, generally admitted to these sacred precincts; but a French gentleman gained access to the Sivaic temple at Treviscare, and there found a granite pedestal in which was a large cleft representing the female sex. On this base was a column supporting a basin, from the center of which arose a colossal linga about three feet in height. This sanctuary is lit from above only.

Figure 198, which is said to be Time and Truth wor-

Fig. 198.

shiping Siva, illustrates this ancient worship in India. In this there is no suggestion of the feminine principle. The Serpent is a common religious symbol in India — as indeed it is nearly everywhere — and is frequently used in connection with the linga to indicate passion, power, vitality, and activity — as well as wisdom, discretion, and use — and, hence, active application for increase, both physical and mental.

The Serpent, with the masculine tail in the feminine mouth, (Figure 144, page 102), typing their active union to perpetuate the race — either with or without an inscription — forming the ring of eternity, is a common symbol in India.

The "Staff of Siva" (Figure 146, page 102), consisting of the upright pillar, with the two entwining serpents, is a constantly recurring symbol.

Figure 199 represents Maia-Devi in a sea of serpents worshiping the linga which she holds in her hands in such a position that she can contemplate at once this emblem and her navel — which to her is, in this meditation, the representation of the navel of Vishnu, or creative power. Devi is also frequently represented with a linga on her head.

Fig. 199.

The Tibetan Buddhists (who are Indian in their religion and practice, and who are less progressive, and, therefore, r e t a i n longer the primitive dogmas and ceremonials) are in the practice of seeking the assistance of the divine, when in danger, by building a "Temple of Peace," as shown in Figure 200. The worshipers bow in silent medita-

Fig. 200.

tion and adoration before it; while the priest calls upon it to protect them from their enemies. It is usually built of clay and plastered with lime and whitewashed.

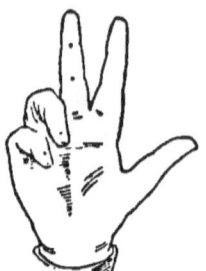

The masculine hand, or hand of wisdom — hand of mystery — is a sign which the Linga-citas interpret as the creative triad. Lingas are made by the women — or by the priests for them — for temporary use, of clay from the Ganges, and offered in Siva's temples, and thrown back into that river after use. The priests of Siva are vowed to the strictest

Fig. 201.

chastity. As they are nude when officiating, any excitement of the imagination which manifested itself in the external organs would be readily noticed by the people, who would proceed to punish such clerical unfaithfulness by immediately stoning the offender.

It is not an uncommon custom for women who are barren to kiss the inert organ of one of these priests, or of an idiot, as a charm to render them fruitful.

Among the Druses, on a certain day, the chief Sheik attends at a sacred place for the purpose of allowing the female devotees, for a similar purpose, to kiss his living symbol of creation.

The Sivaites never carry the linga in procession; and do not present, to the outside observer at least, any indecent ceremonies, or suggest any impurity or indelicacy in the mind of the devotee. They are thus in striking contrast with some of their neighbor phallic worshipers, as well as with Western Orientals, Greeks, Romans, and Egyptians; showing that the use of sexual symbols in worship need not necessarily be associated with impurity of thought or indecency of action in ceremonials.

SACTI WORSHIP.

Indian Mythology teaches a divine masculine creative triad, each of whom have a wife. Brahma's consort is Saraswati, Vishnu has Lakshmi, and Siva, the generator of mankind, has for a spouse Parvati, meaning "mountain born" — referring to the *mons veneris* — womb of nature — or, as she is usually called, Devi. These consorts are known under the general name of Sacti, and are also called Matris — or mothers. Some Hindus prefer to worship the Sacti rather than Siva, just as some pious Christians worship the Virgin, or Holy Mother, more earnestly and more satisfactorily than they do the Father. These worshipers of the feminine are in the East called Sactas.

The worship of the yoni as the emblem of the Sacti is, by its adherents, said to have its authority and origin in the following myth : —

"Siva and Devi, his wife, shortly after their marriage, had a serious dispute about their comparative power and importance in creating new beings. They mutually agreed that each should create a new race of human beings. Siva produced a race who worshiped the masculine deity only. Their intellects were dull, their bodies feeble, their limbs distorted, and their complexion of different shades of color. Devi at the same time created a race who adored the feminine power only ; they were of quick intellect, well shaped, strong, of kindly aspect and had a beautiful complexion. Furious contests ensued between the two faiths, in which the Sactas were victorious. Siva threatened to destroy the victors, but relented upon condition that they forever leave the country."

The Sactas — Yonigas — worship the female emblem or principle with all the devotion that the Lingacitas bestow upon the linga and its interpretations; but with different rites and ceremonies. They interpret Sacti to mean wisdom — it literally means force — thus identifying her with what the Greeks meant by Sophia or Logos, and offer her the most endearing and flattering phrase. She is endowed with lovely attributes and receives very much such adoration as pious and enthusiastic Catholics pay to the Virgin. The ceremonies have, however, another side when the feasting and merry making concludes the ceremonies; then the devotional is replaced by the reveling; the mystic gives place to the real; and the orgies — eating, drinking, and promiscuous mingling of the sexes — may be better imagined than described.

When represented in pictures the Sacti are shown as ordinary women, modestly draped — often with a child in the arms or lap.

The inverted triangle, the circle, the fig, the pomegranate, the sea, all natural concavities — as caves, clefts, fissures, wells, tanks, and generally all that "contains or produces," are symbols or representatives of the Sacti.

Fig. 202.

The Sactas do not use or acknowledge the masculine hand of the triad, but one like that presented in Figure 202, which they call the *Yonic Charm*, or "door of life." This they "look through" to solve all mysteries; that is, they seek to understand the feminine power and principle as the sun of enlightenment. Notwithstanding the facts of former antagonism and wars between the Lingacitas and Yonigas, they are now so tolerant — or so politic, and so curtailed of power — that they are living peaceably side by side as neighbors. They are each a small sect as compared with those who worship both linga and yoni as of the same — or at least each of essential importance and honor as the emblems of a dual or androgynous deity.

SIVA-SACTI WORSHIP.

It must be borne in mind that the last two and the present forms of worship are practiced by a people of similar general character and habits of thought and industry; that these worshipers are mingling more or less freely together; that their peculiar dogmas, ceremonies and symbology are continually approaching and often even coinciding with each other; and that these dogmas, ceremonies and symbols are traditionally as well as esoterically interpreted differently to the initiated and the ignorant. It is, therefore, impossible for an outsider — and especially one of a different race,

9

language, and mental training — to grasp clearly the
subtile distinctions of doctrine, or interpret very certainly the graded differences of interpretation which
they give their ceremonies and symbols. It is, therefore,
probable that the dogmas and practices of one of these
sects may be in some cases attributed to the others.

Part of the Hindus reconcile the two above men-
tioned systems, and quote two myths to explain and
authorize the new departure. One myth is : —

" The divine cause of creation experienced no bliss,
being isolated — alone. He ardently desired a com-
panion; and immediately the desire was gratified. He
caused his body to divide, and become male and female.
They united, and human beings were thus made."

The other allegory says : —

" Siva and Devi found that their mutual concurrence
was essential to produce perfect offspring ; and Vishnu,
at the solicitation of the goddess, effected a reconciliation between them; hence the navel of Vishnu was
worshiped as one with the sacred yoni."

Modern Hindu phallic worship is mainly of this type ;
and its adherents are called Sacteyas. As this sect
unites the doctrines of the other two, it naturally also
combines their emblems. These symbols all, however,
directly suggest, or are interpreted to mean, linga-in-
yoni — that is the masculine and feminine in active
union in the work of generation. Their ceremonies
are such as illustrate this dogma in imagination and
practice.

The linga is generally represented as standing in the yoni. The ways of indicating this are innumerable; but the design shown in Figure 203 will indicate the general outline and character of their most common, as well as their most suggestive, emblem. The rim of the vessel represents the yoni; the upright pillar the linga. The field between them is called the Argha. In this illustration we have what is often

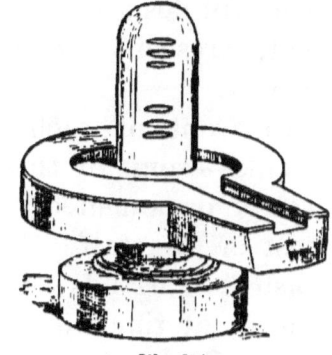

Fig. 203.

not presented; that is, the three bars upon the linga, representing the masculine creative triad; and this again repeated above, which indicates conjunction of the creative powers. The linga, as before remarked,

Fig. 204.

Fig. 205.

is often used in combination with the serpent — to indicate power, passion, and active virility. In Figure 204 we have a more elaborate design, introducing the

linga-in-yoni together with the celestial four — with
cap, and the serpent.   Figure 205 is a copy of a most
beautiful design — a combination of linga-in-yoni, serpent, crescent moon, circles, pentagram, and sacred
fig leaf.   .

In front of each principal temple may be found
a tank — some of them beautifully designed and elaborately ornamented; and in the center of the tank a
mast or flagstaff.   Upon this staff or mast a flag is
hoisted, garlands of flowers are hung, or a light is
placed, at times of special importance.   The temples
of the Sactas have the tank, but no mast.   A high,
but flat elevation, a natural circular or oval depres-
sion, a pond or lake, may often be seen with a pole
or pillar erected near the center.   If a Hindu of this
faith dig a well or build a cistern, he does not consider his work finished until, after appropriate ceremony, on a lucky or sacred day, a mast is inserted
in the center of the mysterious yoni; thus uniting the
original Siva and Devi — in the "marriage of the
linga and yoni."

As before stated, Figure 119 exhibits one of the
most common, and the most sacred, of emblems of
India.   This is the key for interpreting all other symbols.   This same idea is variously expressed, with del-
icate shades of difference, in the symbols numbered
from 120 to 141, pages 98–100, and from 178 to 182,
page 111, all of which are of Hindu origin.

The Sacteyas draw three horizontal lines in black,
and a circle, in red, upon their foreheads, similar to

Figure 126; and consider it a wonderful charm against all evil, as well as a profession of their faith.

Fig. 206.—ARDANARI-ISWARI.
[From an original drawing by Chrisna Swami, Pundit.]

Figure 206 gives Ardanari-Iswari, and is an attempt to express in a design — the following from the Purana : —

" The Supreme Spirit, in the act of creation, became, by Yoga, twofold; the right side was male, the left

was Prakriti. She is of one form with Brahmah. She
is Maia, eternal and imperishable, such as the spirit,
such is the inherent energy (the Sacti), as the faculty
of burning is inherent in fire.''

This design is, however, much conventionalized from
the original; for where the *Crux Ansata* appears in our
reproduction, the original shows, in realistic detail, the
living and erected "linga-in-yoni."

In Figure 207 is reproduced one of the most elaborate, as well as one of the most beautiful, designs,
both in execution and interpretation, that is to be
found in connection with this worship in India. The
religious teachers say : —

"When one can interpret this emblem of the androgynous divinity, he knows all that is known; and
that to learn more he must be enlightened to read yet
more mystically the inexhaustible truth incarnated in
this most wonderful symbol."

This picture has been commented on by nearly
every student of Hindu religion, in all degrees of spirit,
from scorn to rapture.

Figure 146, page 102, is a symbol common to the
Sacteyas, who interpret it as the linga entwined by a
male and female serpent in sexual congress. This idea
is more realistically represented, on certain occasions of
high religious ceremonies, by the women, in grand
procession, carrying, between two living serpents, a
gigantic linga, decked in ribbons and flowers, the
prepucial end of which they present to an equally
prominent yoni. They likewise use the symbol of a

serpent with its tail in its mouth, Figure 144, as representing a perpetuation of the race through the creative activity of the sexes. They also use the design of the

Fig. 207. — ADDHA-NARI.

chest or ark, in which the serpent, or passion, is supposed to be alive — but dormant, as a symbol of Devi. The Nagas pray that the serpent may come out of the ark — passion be aroused, sexual union be thereby consummated — with the blessed result of many and worthy children. In Maia worshiping the linga,

Figure 199, they recognize Devi — herself the feminine creator, and, therefore, worthy of worship — as recognizing her masculine consort as divine, and thus directing her adorers to also recognize and worship the linga and all it is interpreted to represent.

THE TORTOISE is an important emblem in the Hindu mythology. They represent the world resting upon an elephant supported by a tortoise. . It was chosen because it is popularly supposed to be androgynous, on account of its great tenacity of life and its great fecundity.

 The frequency and rapidity with which it protrudes its head from its shell and with-draws it, chang-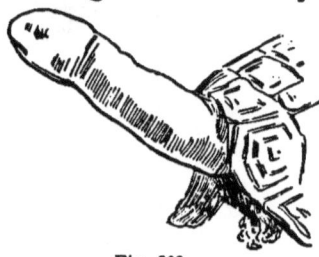

Fig. 208.   Fig. 209.

ing from an appearance of repose to one of energy and action, as well as the configuration of its head and neck when aroused, would readily suggest to the mystic Hindu — the acting linga; while a front view would equally bring to his imagination the sacred eye, or arba-il.

THE LOTUS.— Brahma is represented as sitting upon his lotus throne. The lotus was the most sacred flower among the ancients, and to them typed the two powers of generation. The germ symbolized the linga, the filaments and petals the yoni. The lotus is a *nymphœa*. Nympha signifies a young nubile woman, a certain part of the yoni, and the calix of the rose. Hence, a maiden is symbolized as being, or having, a rose. The lotus not only signifies the andogynous creator, but typifies Sacti.

The modern Hindu phallic worship which recognizes the essential importance of both the sexual elements in generation is usually spoken of as Sacteyan worship, in much the same way that in the West all kinds of sex worship is called phallic worship. All Sacteyan worship requires the use of some or all of the five following necessities: flesh, fish, wine, woman, and certain mystical performances called dancing, but which, unlike the dances of the West, consists of a pantomime made up of dramatic action, gestures, twistings, and undulatory and expressive motions of the arms, legs, and whole body. This dancing is at once poetical, sensuous and skillful, and is performed by professional nautch girls. Every temple — of this faith — of any note in India has a troop of these nautch girls. They are generally selected, by the priests, when quite young on account of their beauty, health, strength and activity. From infancy they are trained in dancing, vocal and instrumental music; and at an early age initiated into all the mysteries and duties of their profession. Their natural beauty is heightened by all the accessories of drapery, jewels, seductive arts, and general feminine witchery. Their chief ostensible employment is to chant the sacred hymns and perform nautches before the idol at high festivals. But they have another office to perform. They are the acknowledged mistresses of the officiating priests, and are required to prostitute themselves — in the courts of the temples — to all who desire and will pay for their possession, and thus secure funds to sustain and enrich the temple to which they are attached. As they are beautiful and accomplished in all seductive and passion-arousing arts, healthy, and, therefore, safe companions, and as it is considered honorable on their part as well as in their patrons thus to swell the temple revenue, and as there is absolute secrecy as to their patrons, it need not be wondered at that they are much sought after, and well paid for this part of their service.

A similar class of women are found in many other parts of Asia; and it is said they are far from rare in Turkey.

These "votaries of the deity," "women of the idol," "Devadasi," "women given to God," are looked upon as holy devotees of the faith. Any woman, however, who prostitutes herself for selfish gain in India is an outcast who bears a disgraceful name.

The principal ceremonies include the worship of power, and require the presence of a young, beautiful, and naked girl as the living representative of the goddess. This girl is generally selected from the nautch company; and the one chosen esteems it as an especial honor, as a tribute to her beauty, accomplishments and ability. The peculiar duties of this office, the nautch girl is, by experience, every way fitted to meet with better grace and more satisfaction than an innocent and unsophisticated girl. To this naked girl meat and wine are offered, and then distributed among the wor-

shipers.   This is followed by the chanting of sacred
texts and dancing.   The celebration ends with an orgy
of the most licentious character.   The woman who
in this ceremony takes the part is ever afterwards called
Yogini — attached, which is equivalent to a secular
nun — and she is ever afterwards supported by alms.
Although all parties engage in this worship — of
course as a religious ceremony pleasing to the divine —
yet the women who are, or claim to be, faithful wives
are warned not to associate with one who has thus
officiated as a representative of Sacti.

Sacti is personified as the deified *vulva;* and in adoring her mentally the worshiper imagines a yoni, in
which he tries to see a chapel, which he is to enter, and
in which he is to worship.

The members of this sect who participate in this
Sacti-puja initiation are sworn to secrecy.   Gradually,
however, those who are initiated become less reserved
as to the fact of their initiation into the mysteries;
but the mysteries and the forms of initiation are not
revealed.

The sect known as Kauchiluas are near akin to the
Sacteyans; but are distinguished by a peculiar rite,
which " throws into confusion all the ties of female
relationship."   Natural restraints are wholly obliter-
ated for the time being, for a community of sexual
partners.   The women — matrons and maids — deposit
their bodices in a box — each woman and bodice being
numbered by the priest.   At the close of the ceremonies each male worshiper takes a bodice from the

box, and the woman who has the same number found on the garment — even were she sister or daughter of the man who draws it — is his partner for the night in the lascivious orgies that follow. All these ceremonies, in their wildest excesses, are engaged in by the most devout and pure-minded men and women — most of whom, outside of this ceremony, that they consider a sacred and solemn obedience to their religious requirements, are, according to their ideas of purity, as modest and chaste as any devotee of their more enlightened neighbors of the Western civilization.

A peculiar custom, still common in India, is thus described by General Furlong : —

"Many a day have I stood, at early dawn, in the door of my tent, pitched in a sacred grove, and gazed at the little group of females stealthily emerging from the adjoining half sleeping village, each with a little garland or bunch of sweet flowers, and perhaps costly oil, wending their way to that temple in the grove or garden of the god and goddess of creation ; and, when none were thought to see, accompanying their earnest prayer for pooli-palam (child-fruit) with a respectful abrasion of a certain part of their person on linga-jee, and a little application of the drippings that are forever trickling from the orifice of the Argha."

In Oriental villages it is common to see two stones — one circular, and the other small, smooth and upright — near together ; they indicate the male and female. Women step upon the circular stone, adjust their drapery so that perfect contact with the vulva can be assured, and seat themselves upon the upright stone, with at least partial entrance — repeating a short prayer for any desired favor.

According to some Hindu sects women of or above the age of puberty, who are maidens, cannot enter Paradise. They, therefore, if denied marriage, rupture their hymen by means of an idol with an iron or stone linga. Brides in Pondicherry sacrifice their maidenhood in a similar way — in honor of the deity — to whom they first belong. This was not an unusual custom in many ancient nations. The Moabitish maidens always thus sacrificed their maidenhood, as a religious duty, to their deity, Peor, before becoming kedesha among the Jews.

Some Hindu women of some sects regard a child resulting from intercourse with a peculiarly saintly priest as an incarnation by Vishnu; and, if they can agree upon terms, the official will generally accommodate her.

PHALLISM IN EGYPT.

The oldest and dimmest traditions, the earliest writers, and the remains of the most ancient sculptures, tell us of phallic dogmas, ceremonies, and symbols being abundantly general in Egypt. In the ancient Egyptian religion, the good and creative power — the masculine principle — the active principle, as they generally called it — was attributed to, or incarnated in, Osiris. Osiris was the child of Time and Matter. He was worshiped as the being who dwelt invisibly in the sun; so the sun was one of his emblems. From this idea of the sun, and its heat and light as creative powers, he was also represented by fire — celestial fire; and, hence, by the upright triangle — which is a symbol of Osiris, because it is a symbol of fire. The bull was, however, his chief symbol, and was regarded as his real self, incarnated in living form. This sacred bull was said to be miraculously begotten by a ray from heaven, and bore certain marks which revealed his divine parentage. The worship of the bull was, in later times, connected with the constellation Taurus in the Zodiac; but this was a later adaptation, and the probability is that the constellation was so named by those who "adapted" the union of the two cults. In all interpretations it must be borne in mind that time-honored symbols, as well as sacred days and seasons, are persistently retained — for the masses prize forms, times, and ceremonies. The hawk was also a representation of Osiris as an emblem of directing power. The Nile, upon which depended their crops, was called by the Egyptians the outpouring of Osiris, so when they personified the Nile or any other river it was represented as a bull — or with the attributes of that sacred animal. In short, all beneficent and productive moisture was venerated as being the substance of the semen of Osiris. By intercourse with Isis he produced all living beings. He was reported dormant or absent for forty days in each year — which was a season of sorrow and lamentation; and his body was said to be repeatedly torn in pieces by his bad brother, Typhon.

The goat was one of the sacred animals of Egypt, and, probably on account of its well known salacious peculiarities, was worshiped as the personification of the masculine principle — or male creator. It seems, however, that the goat, both male and female, were used in a more sensual sense — to type the divine powers as exhibited in human manifestation — hence, human virility, passion, and its satisfaction and fruit. A part of the veneration bestowed upon this animal at Mendes, which was especially celebrated as the great center of Caprine worship, was for the women to offer themselves sexually to the goat. This unnatural copulation, Herodotus tells us, the goat accepted, and the union took place publicly in the assembly. The female goat was also sacred, but not so highly esteemed, or at least not so generally made prominent in the ceremonies or in symbolic art representations. Still the women did not monopolize the practice of caprine copulation, as is shown by occasional references, and not infrequent sculptures and paintings representing men in sexual union with female goats.

This orgy was well calculated to suggest, even if it did not produce, the satyrs and fauns — which play such an important part in Grecian mythology; and by arousing the hopes, quickening the imagination, and exalting the passions, it was well calculated to render prolific the women who took part in or witnessed the ceremony.

The augurs who prompted the oracles of Juno, when consulted as to the cause and remedy of barrenness among the Roman women, probably wished to introduce this practice when the response was: "Let the rough goats approach the Trojan matrons." But this mandate was executed in the very different way of sacrificing the goat and cutting the skin into thongs, with which the women were scourged upon their bare backs. The desired result of child-bearing was, however, thus attained, showing the powerful effect of flagellation and an exalted imagination; for Ovid tells us "speedily was the man a father, and the wife a mother."

This sacred goat of Mendes was by the Greeks transformed into their god, Pan, and represented by a personification half goat and half man. Satyrs and fauns seem to be degenerate and purely sensual derivatives from Pan.

Representations of Pan, in some instances, show him with rigid and strained muscles, his face wild with passion, and his generative organ ready for his characteristic work. He is at other times shown with relaxed muscles and a jaded countenance, as if wearied by his depleting excesses; in all cases, however, his phallus is of exaggerated proportions, thus representing his peculiar characteristic.

The hereditary priests of Egypt were, when advanced to the sacerdotal rank, first initiated into the mysteries of the goat, as a preparation for the higher and more divine mysteries of Isis.

The mysteries of the goat, and the sublimer arcana of Isis, as in fact all the esoteric interpretations of the

Egyptian cult, was a sacred trust which was known only to the initiated priesthood (and some secrets were imparted to only a chosen few of the most enlightened and most trusted priests), and was guarded so zealously and successfully that little is known concerning them. While their religion was clearly phallic — recognizing both masculine and feminine creative deities and the necessity of their sexual union in producing new beings, and while these views were very realistically represented in their religious ceremonies, still the worship — or, at least, the "mysteries" — of the feminine were the more exalted.

In later times, the goat was an important element in the initiations, ceremonies, and occult work of the Templars.

But the Templars, in introducing the Goat of Mendes, and in the inauguration and continuation of their sabat, were only adapting to their use a well-known ancient, effective and occult ceremony —which, to the instructed and intelligent initiate, had a holy esoteric interpretation, and which was well calculated to test, secure, and maintain the neophyte's integrity, endurance, and enlightenment.

The obscene sabat of the sorcerers bore the same relationship to the Templar ceremonies that prostitution does to holy wedlock.

The Templars, by a series of impressive and instructive ceremonies, sought to teach transcendent truths, which, being contrary to the dogmas of the church, were unsafe to teach openly. For this reason the neophyte was

10

severely tested and rigidly vowed to secrecy. The profane sabat, or, as it was called, the "witches sabat," was practiced by those who mistook the shadow for the substance, and who engaged in the wild orgies — not for enlightenment — but for selfish gain or lustful gratification — and were secret because they were criminal.

Osiris was represented as a man with an enormous movable phallus, to signify the prolific procreative power of the good generative principle. He was sometimes represented with three phalli, to symbolize his active creative energy in the three elemental worlds — air, earth, and water. The women carried these manikins in their sacred processions in some of their religious ceremonies.

Typhon was the personification of the evil power or destroyer, and was represented by the Hippopotamus — the most savage animal known to the Egyptians. He was also represented by material fire. To show the final power of good over evil, it is said Horus castrated Typhon, and there are statues of the former with the phallus of the latter in his left hand.

The same idea is found also in the Hindu cults, from which it was probably adopted by the Egyptians, and also in the Grecian myths, which were borrowed from one of the above two sources. Saturn is represented as having cut off the genitals of his father. In ancient times a castrated god — and, therefore, a non-generating eunuch — lost all claims to divinity. Defeat in any contest might be condoned, or the vanquished once be the conquerer next time ; but lost generative organs must be restored, or the deity was repudiated. This restoration was said to be often accomplished — but that peculiar surgical operation is not now understood.

There may be seen even at the present day on the walls of the ancient temples at Karnak and Thebes, as well as in the temple at Danclesa (which was built much later, but in imitation of the ancient Egyptian art), many phallic designs, which illustrate how intimately the ideas of virility and religion were interwoven in the old Egyptian civilization. There are many figures of their gods and kings showing them in manly proportions, and their abundant creative energy or virile power indicated by the erected penis.

On the other hand, in the scenes which commemorate victories over their enemies, they are represented as returning in triumph, with multitudes of captives, many of whom are shown as undergoing the mutilation of castration ; and there is seen, in one corner of the picture, heaps of the genital members which have been cut off from these unfortunate captives.

Asiatics and Aryans, ancient and modern, counted the heads of those slain piled up before them. The Africans of olden time, like their dusky representatives of the present day, do not count heads, but enumerate the genitals removed from their captured enemies.

The former gratified a temporary revenge, and buried or gave to the crows the dead bodies of the vanquished. The latter took a more lasting triumph, and utilized the emasculated captives who had a producing

or market value as slaves. This was a practice some-
what in use among the Jews (whether justified or only
tolerated we need not discuss).

The slave trade of Africa, which furnishes Turkey
and other localities with eunuchs in modern times, is
simply the remnant of this ancient custom. Typhon
is said to have destroyed one of Horus'. eyes, so a certain order of Egyptian priests were deprived of one
eye — in commemoration of this mutilation of their
deity. Many of the Egyptian priests and priestesses
who appeared in Rome were thus deformed.

The good feminine creative power — passive, receptive, and nourishing — was personified in Isis. This
character was still more generalized, so as to include
universal nature. She says : —

"I am nature, the parent of things, the sovereign of
the elements, the primary progeny of time, the most
exalted of the deities, the first of the heavenly gods and
goddesses, the queen of the shades, whose single deity
the whole world venerates — in many forms, with
various rites — under many names. The wise and good
Egyptians worship me as Isis."

Isis is identified with nature — hence, with the earth
and with the moon. Her representations are innumerable, but the cow, either as a mere animal or as a young
and finely formed woman with a cow's head, is the original and most sacred symbol. She is also represented
as a woman with a child — Horus — in her lap, or
standing by her side, his mouth at her breast. Figure
210 gives one of these pictures, which is very sug-

gestive of the Assyrian "grove," portraying to the initiated the "door of life" through which every human being enters the world. The whole design shows Isis and Horus in "the door of life," while the bells indicate the breasts, multiplied in number and size, so they are sufficient to abundantly nourish all whom the Door of Life ushers into existence. The bells — thirteen in number — are explained very differently in the Assyrian cult; but the phallic character is always maintained.

Fig. 210.

The sun over head — which is a symbol found over the porticos of many Egyptian temples — signifies the central sun — the masculine creative power —Osiris. The crescent moon is again the feminine — the virgin — the mother — Isis. The position of the sun and moon together is also creation — sexual union — marriage of Isis and Osiris.

Notwithstanding Isis is the Divine Mother of Horus — that is, of all created beings and things — and that this motherhood is the natural result of copulation with Osiris, still she is worshiped as the Celestial and the Eternal Virgin, who, by the use of her Sistrum or Virginal Magic Wand, drove away Typhon — or evil, from her presence. This Sistrum, shown in Figures 211-213 represents the yoni, thrice barred across — thus closing the Door of Life. The bars are also bent so they cannot be removed except by the "*Celestial Magic Wand.*" The Virginity of Isis — the Celestial

Mother — was a tenet of the Egyptian faith at least fifteen centuries before the Virgin Mary bore Jesus.

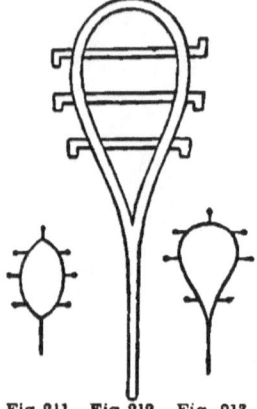

The Egyptians symbolized their divine triad by a simple triangle. They compare the perpendicular to the male, the base to the female, the sides to the offspring of the two creative powers — Osiris as the beginning, Isis as the medium or receptacle, and Horus as the accomplishing. The pyramid — the ancient and modern achievement and wonder

Fig. 211.   Fig. 212.   Fig. 213.   of Egypt — is the solid triangle; each face a triangle, the base and four faces — again the " Four Great Gods."

Vivant Denon found at Thebes the mummy of a woman who had probably been a lady of rank. In the vagina of this mummy there was inserted the embalmed phallus of a bull, which had, in all probability, been taken from a sacred animal after his death. It was then embalmed and placed in its human receptacle as a charm against evil spirits which the ancients believed tormented the souls of the dead.

The Greeks and Romans frequently placed figures of the phallus in tombs from similar motives.

Josephus tells us that the custom of saying grace before meals was practiced by the Egyptians; and when seventy-two elders were invited by Ptolemy Philadelphus to sup at the palace, Nicanor requested Eleazar to say grace for his countrymen, instead of those Egyptians to whom that duty was committed on other occasions. In short, they were punctilious and scrupulous in their observance of the religious ceremonies. These ceremonies performed, they were convivial, hilarious, uproarious, and frequently drunken and licentious — just like modern Europeans and Americans. They believed in the transmigration of the human soul — which they taught passed in its progress through many animals, returning again as man in about 10,000 years for ordinary men; but in about 3,000 years for the good and wise.

Herodotus says the Egyptians were the first people to assert the immortality of the human soul.

PHALLISM IN ASSYRIA, PHŒNICIA, SYRIA, AND PHRYGIA.

The worship of the Assyrians — including the Babylonians, the Phœnicians, the Syrians and Phrygians — was the same in essence and nearly the same in general character. Their deities bore different names, and were, in the different civilizations and times, regarded as having some peculiar differences of characteristics and powers; and were represented under different forms. They were always, however, distinctly and intensely sexual, vitally and actively virile. The human organs of generation were their constant and especially cherished symbols. And their worship always included ceremonies in which the devotees enthusiastically engaged in the creative activity of striving to imitate —

without any hope of ever equaling — the propagating performances of their deities.

The supreme masculine creator was by the Assyrians called Bel; and manifested in the male triad, Asher — after whom the empire was named, Anu and Hoa. By the Phœnicians he was called Baal; by the Phrygians, Atys; and by the Syrians, Adonis. The feminine consort of Bel was Mylitta, also called Ishtar. The Phœnicians named her Ashtoreth, or Derceto, and represented her as a woman terminating from the hips down in a fish. The Syrian goddess was also Derceto, but, unlike her Phœnician namesake, was a complete and voluptuous woman; who was, however, sometimes represented as a fish to symbolize her fecundity. She was also called Atargatis, and as such shared honors with her bastard daughter, Semiramis, who was represented by a dove; because the cooing of the dove in the night sounded like the Syrian word which meant coition. Cybele — also known as the mother of orgies — was the Phrygian goddess.

These deities were generally thought of and represented as distinctly sexed masculine or feminine beings. They were, however, often worshiped and figured, realistically and symbolically, as androgynes. It is probable that at a later period these deities were generally known — in addition to their local names — as Jupiter, Juno, and the "mysterious third." Just what this "mysterious third" meant was an esoteric and carefully guarded secret, revealed only to the specially favorite associates of the inner circle of the priesthood. It has been variously explained as the creative act of the divine creators, the children as the result of this act, and as the illuminated prophets who talked with the gods and then instructed the people. The highest interpretation was " Love — divine impulse to create."

While the dogmas of these countries named the masculine and feminine deities together, and taught their equal importance and honor, there were some very curious practical outworkings. The temples were built to the goddesses. The male emblems were often very realistic, and always numerous. The priests and principal temple attendants were males or eunuchs, while the worship paid was principally to the feminine deity. The men directed the rites and ceremonies, yet the women were the more enthusiastic worshipers. While virginity and chastity were there, as elsewhere, woman's greatest treasure, and profane loss of them was punishable with death, still they enthusiastically sacrificed both — the men gladly consenting — in religious orgies in honor of their celestial virgin mother. Women who at home and in society were modest, chaste, and honorable, when worshiping engaged passionately in the wildest sexual excesses, and even in the grossest and most unnatural satisfaction of frenzied sexual passions.

The religion — and consequent ceremonies — of Syria and Phrygia was at one time very peculiar; it was broadly and intensely phallic, and ran to the extreme of sexual symbolism and licentious excess among the masses of the worshipers, while it, at the same time, required emasculation of the priesthood and temple attendants.

Lucian describes the Syrian temple and worship at great length, and in wonderful detail — using, however, the Greek names for the deities instead of their local equivalents. The following is an outline of his statements : —

" The magnificent temple of Atargatis, at Hierapolis, is situated on a commanding eminence in the midst of the city, and surrounded by a double wall. The porch of the temple is two hundred yards in circumference. Within this porch, in front of the temple, are two enormous phalli, each a hundred and fifty yards high, and bearing the inscription, 'These phalli, I, Bacchus, dedicate to my step-mother, Juno.' A man once a year ascends to the top of these phalli, remaining there seven days. At the right of the temple is a little brazen man with an enormous erect phallus. Outside the temple there is a very large brazen altar and a thousand brazen statues of gods and heroes, priests and kings. The temple, into which any one may go, has golden doors, a roof of the same material, and the interior is gorgeously garnished with a blaze of golden ornaments. It is filled with a heavy and delicious perfume which clings a long time to the visitors' garments. On the left as one enters there is the throne of the sun, but no representation of that luminary; because, they say, all may see the sun himself, and, therefore, need no symbol. There is also the statue of a woman in man's dress. Next is the statue of Apollo, with a long beard and clothed. All the other statues are nude.

Next are the statues of Atlas, Mercury, and Lucina. Within the temple is a sanctum, which is entered only by the high priest and his most holy associates. In this sanctum are golden statues of Juno and Jupiter — which the priests call by other names. The latter is seated upon a platform supported by bulls. Juno is seated on a like stage borne by lions. In one hand she holds a sceptre, in the other a distaff. Her head is crowned with rays and a tower. Her dress is profusely adorned with gold and precious stones of all kinds, brought and presented by devotees from Egypt, India, Ethiopia, Media, Persia, Armenia, and Babylon. Between these two statues is the reverenced but unnamed 'mysterious third.' There were over three hundred priests attached to this temple, some of whom kill the sacrifice, others carry the drink offerings, others are fire bearers, while the remainder wait on the altars. They all wear white garments and a peculiar felt cap. They each year elect one of their number as high priest, who is, during his term of office, clad in purple and wears a golden tiara. These priests are all self-castrated. Attached to the temple are also crowds of other persons — musicians, galli or sodomites, and enthusiastic or fanatic women. All these attendants come to the temple to the sacrifice, which occurs twice a day. A peculiarity of their service is that they make offering to Jupiter (Adonis) in silence, while their sacrifice to Juno (Atargatis) is accompanied with music — for which no reason is given to the uninitiated."

The method and occasion of the self-castration of the eunuch attendants and of the candidates for priesthood was in all respects like the same ceremony among the Phrygians, which is described by a learned French author, in substance, as follows : —

" Once each year in the springtime there was a wild and noisy, though a sacred and solemn, festival. It began in quiet and sorrow, for the death-like sleep of Atys. On the third day joy breaks forth, and is manifested by delirious hilarity. The frenzied priests of Cybele rush about in bands, with haggard eyes and disordered hair, like drunken revelers and insane women. In one hand they carry burning fire brands, in the other they brandish the sacred knife. They dash into the woods and valleys, and climb the mountain heights, keeping up a horrible noise and continual groaning. An intoxicating drink has rendered them wild. They beat each other with the chains they carry. When they draw blood upon others or upon themselves they dance with wild and tumultuous gesticulations, flagellating their backs, piercing their limbs, and even their bodies. Finally, in honor of the god they worship, they turn the sacred knife upon their genitals and call upon their deity, showing her their gaping wounds and offering her the bleeding spoils of their destroyed virility. When they recover from this self-inflicted unmanning, these eunuchs — or, as they call themselves, galli — adopt woman's dress. They are then ready to become priests, or, failing in that, to take their place as attendants of the temple worship ; or to engage in pederasty for the benefit of the temple treasury, whenever their patrons prefer such indulgence to ordinary fornication with the enthusiastic women."

While this fanatical — but, to the participants, awfully solemn — procedure of the would-be priests and temple servitors was taking place on the hills and in the valleys, a very different ceremony was performed in or near the temple. There the orgy was as wild, but less bloody, and more licentious, but equally phallic. The sexual rites were of three orders: First, the devotees could choose sexual association with the "temple women," who were available to whoever desired to pay for their service — the sums thus realized being turned into the temple treasury; or they could, if they so desired, engage in what Paul describes as "women changing the natural use into that which is against nature; and likewise also the men, leaving the natural use of the woman, burned in their lust toward one another; men with men working unseemliness." The galli attendants at the temples were also sodomites, and the price of their uncleanness increased the income of the temple. Those who did not care to engage in these rites could, under certain rules, join each other in fornication; and, in many instances, all bonds of blood or kinship were totally ignored.

The character of the religious services in Babylon is shown from the fact that the chief temple in that city was called by the name of Bit-Shaggathu, which means literally "the temple for copulation."

Besides many other phallic ceremonies, every native woman in Babylon was obliged, as an imperative religious duty, to present herself in the temple of Mylitta, and, once in her life, deliver herself to a stranger. They came to the temple wearing a crown of cord about the head. Most of them were seated in such a manner that those desiring their company could pass along straight aisles among them; thus securing a full and fair view of the candidates. Some, however, proud of their wealth and rank, came in covered carriages, attended by servants, and remained thus apart. Whenever a woman thus presented herself, she was expected to be in constant attendance until she attained the object of her visit. As the stranger passed along the aisles, having made his choice, he threw the selected one a piece of silver, saying: " I beseech the goddess Mylitta to favor thee." No matter what the value of the silver, large or small, she must accept it from the first to offer it: for it was thus made sacred and applied to religious purposes. She then followed him outside the temple to one of the semi-seclusive alcoves provided for the purpose, and there had sexual intercourse with him. Having thus performed her religious devotions to her goddess, Mylitta, she returned home, believing she was purified. Any subsequent deviation from chastity would be considered mortal sin.

Many were continually coming to thus present themselves in the temple; and, of course, many retiring after their devotions.

It will readily be seen that those endowed with beauty of features or symmetry and richness of form were not long detained, for no refusal was allowed; while the unattractive or deformed were often compelled to experience a weary waiting.

Similar customs were followed in Armenia, Cyprus, and in fact in most ancient nations in some period of their religious development. This practice, however,

must not be confounded, as it often has been by un-
careful writers, with "consecrated prostitution," spoken
of elsewhere.

In connection with the worship of Assyrians and of
the neighboring nations which they influenced, there
occurs a very remarkable, as well as a very elaborate,
symbol. It is of very frequent occurrence in the sculptures of Nineveh. It is called by the name of Asherah,
which, in the King James version of the Bible, is translated "groves," and is, therefore, of special interest to
the Christian world.

Dr. Inman, in his Ancient Faiths, identifies the
"Asherah" with the female "door of
life." He says: "The Asherah, or
grove, Figure 214, shows a central fissure — the door of life. This is barred
more elaborately than the sistrum shown
in Figure 212, but with the same significance. Above the fissure is a fan-like
emblem, representing the clitoris — divided into seven parts, which represent

Fig. 214.

the seven planets, or the seven days in the moon's
phases. Around the fissure is a fringe, as in nature,
which is artistically arranged in tufts or curled braids.
These are thirteen in number, indicating the number of
fertile periods in a woman's life each year. In Figure
215, of which the "grove" is the central object, the
periods are also found by counting the tufts on each
side, the one at the top being common to both and
forming the thirteenth.

Mr. Newton — an equally erudite student — gives it
a more elaborate interpretation, which is, however,
quite as phallic. The truth probably is, that when
used, it was successively — and, perhaps, contemporaneously — interpreted both ways, by those whose views
of the relative superiority or equally exalted value of
the male and female principles called for the special
meaning they gave it. According to the latter writer,
it embodied, in a more complex and veiled way, all that
is contained in the interpretation of the *Crux Ansata* —
or both sexes and their united activity in creation.

The design in Figure 215 shows the grove receiving
the worship of the king and his son or successor and
their attendant genii — their rank and character being

Fig. 215.

shown by their head-dresses, costumes, and the symbols carried in their hands. The kings present to the
grove the " phallic right hand," the symbol of life and
good fortune. They each carry in the other hand a rod
of life or sceptre. The attendants, each with the right
hand, presents the masculine emblem of the pine cone,
and carries in the left hand a bag or basket, in which is

symbolically stored abundance of energy. The winged figure above the grove — originally the dove — is the celestial bowman, with string, bow and quiver full of arrows; which are for the use of all who desire divine vitality and activity in the sensual manifestation of worshiping the grove.

There are numerous representations of the grove and its adoration in many modified forms and combinations; but they all agree in the general character above described. Always the central "door" barred and fringed; always the worshipers — kings, divine beings, warriors, or laymen, offer gifts of phallic and creative import. The homage took generally — and probably always — the form of actual copulation among the worshipers.

This grove was evidently the symbol of Ashtoreth, or of the creative union of Baal and Ashtoreth. The practical ultimation in this service took place between the male and female devotees, who retired to a small bower, or arched tent, called a *qubbah* — which is also the Hebrew name of the yoni. Each kadeshah had such a tent attached to or near the temple or worshiping place where homage was paid to the " grove."

Many statuettes found in Nineveh, unquestionably represent the feminine deity, as the yoni is very obtrusively represented — the hair on the *mons veneris* being conventionally curled, after the manner of the beards of the males in ancient Assyrian statues. In others, the fissure and hirsute appendages are entirely omitted. No explanation is known for the difference.

11

The royal collar, here presented, was a common jewel in Babylon, Assyria, and Rome. It was worn by all classes in those countries, and is worn now by pious worshipers of Maha Deva, in India. On the left is the ever-recurring masculine triad, representing the Divine Father, while on the right is the crescent moon,  the symbol of the equally exalted feminine creatress — the eternal Virgin-Mother. The horned cap,

Fig. 216.

next to the trident, is the signature of royalty — or of the divine man — the acting creator or "word." The cross here again represents coöperative activity of the divine creators in generating humanity — the Arba-il — the divine four — populating the world. The double triangle in the circle, with the center marked, is a summing up of all creative arcana. It is sexual union; it is Siva and Sacti — Jupiter and Venus — the sun and the moon — the divine descending into the human, which rises to receive the celestial. In a word, it is the generation of new creatures — on whatever plane the beholder occupies — and according to whatever love and wisdom the translator acknowledges.

Figure 217, from Lajard, represents an act of worship in the presence of the triune representation of the masculine and feminine creators. We have here the celestial, sun, and moon; the mundane, palm tree, and barred vulva — virginity; and the sensual, cone, and lozenge — the fleshy organs. Diana, of the Ephesians, was represented by nearly every symbol of Isis, in

Egypt. She was also shown with a phallic radii — indicating her universal generative power; also as a woman with many breasts — to symbolize her as the goddess of nutrition. Her worship was akin to that of

Fig. 217.

Isis, in Egypt, and to that of Venus, in Greece and Rome. Like the adoration of Anaitis, in Armenia, it was accompanied by the defloration of nubile women and other licentious ceremonies. Isa or Disa, the Scandinavian goddess, was represented — only more rudely — in the same manner as Diana ; also as a pyramid surmounted by a cross and circle. Reindeer were sacrificed to her, and their testicles hung about the neck of her statue.

This highly mystic design is copied from a Babylonian gem figured by Lajard, and is an illustration of how fully and clearly the Assyrians understood, and how forcibly and tersely they expressed the ideas of phallic worship. The palm tree, or "tree of life," represents the phallus with all its interpretations, and in this case probably indicates the great or universal creative

Fig. 218.

power, principle, or person — depending for its special meaning upon the intent and intelligence of the one who translates the scene. The tall stamen,

with the two fruits, one at either side of the base,
symbolizes the phallus and tests; while the ovals on
either side of the upper point indicate the yoni, with
all the occult significations of those organs. The
animals — spotted goats suggestive of great sexual
power and fecundity — rampant, represent passion or
desire. The crescent moon of Isis, over the head of
the male goat, symbolizes the feminine creative power;
and the lozenge below and in front points to its physical manifestation in sex. The wings tell of interpretation, while the erect phallus shows readiness and
power in physical activity. The crescent moon, on
the female goat, near the tail, shows the sex and desire,
while the *fleur de lys* — emblem of the masculine
triad — below and in front, suggests its satisfaction.
The priest, who is androgynous — shown by the peculiarity of the skirt — points to the central palm tree,
explaining and urging its worship, and the consequent
obedience to its teachings — physical and mystic. Of
course, he is not teaching animals, but virile — and,
therefore, exemplary — men and women, who, in the
condition suggested, can be more modestly represented
by the rampant and prolific goats.

The accompanying design, copied from *Lajard's
Researches sur le Culte de Venus*, and taken by that
author from an ancient gem, was originally engraved
upon the lower face of a cone-shaped white agate.
White stones — and particularly agates — were much
prized as emblems; and with the name or symbol of a
favorite deity cut upon them were especially sacred.

In the Apocalypse the promise is : " To him that over-
cometh I will give to eat of the hidden manna, and I
will give him a white stone, and upon the stone a new
name written," which was, in all probability, " mine
own new name," referred to in
the next chapter, as written upon
the same person — he that over-
cometh.   This new name of the
" Faithful and True — the Word
of God," is subsequently
given — written " upon his garment, and on his ' *thigh*,' KING
OF KINGS, AND LORD OF LORDS."
The cone is the Sivaic symbol

Fig. 220.

of the phallus or masculine generator.   It is also sacred
to and the emblem of Venus — not the Grecian
Venus of desire or passion — but the androgynous
deity, or bearded Venus Mylitta.

On the right is a bare feminine face, on the left a
bearded masculine face, and the two heads are united
by the inverted triangle or feminine symbol, and surmounted by the radiating solar corona or masculine
sign.   Across the bust are masculine girdles, below
which appears the inverted triangle again feminine.
The bare masculine arm, and the feminine arm shown
by the bracelet, and the peculiar form of the drapery —
the upright lines in the center and the drooping lines
on either side — from the waist downward to the feet
carry out the same dual symbolism and again suggests
the Apocalyptic androgynous " Son of Man clothed

with a garment down to his feet, and girt about the
breasts (the word, in the original translated breasts,
being — not that which indicates the masculine bosom —
but the feminine, the nutritive, and milk-giving breast)
with a golden girdle." Over the head is a triad of six-
rayed stars — the conjunction of the masculine and
feminine in generation. The crescent moon of Isis is
above, and the feminine cup below, the female snake on
the right. A male serpent — shown by its slimmer and
sharper head — spreads its wings as if attacking this
female. The six-rayed phallic star, the points meeting
in a circle, is in perfect harmony with the whole design.
The male serpent on the left is approached by a winged
and aroused female of its kind. Below the serpent is
a phallic vase with a cup over it — the still favorite
form of oil and water vessels in the temples of Siva.
The lozenge or feminine symbol near the male
serpent again indicates conjunction of the sexes — or
dual creative powers. In this little design, therefore,
may be found the whole doctrine of phallic worship:
the masculine creator, whether organ or power —
the feminine creatress, whether organ or principle —
their mutual desire or attraction — their coöperative
activity in the work of generation — and the essential
unity of these organs or powers — in a word, the androgynous character of the great and essential creator.
And surely, the intelligent and aspiring Christian can,
by spiritually interpreting this unique design, read into
it all the transcendental truths of his beloved faith.

PHALLISM AMONG THE JEWS.

Even a casual examination of this subject will establish the fact that phallic worship was known, and many of its rites practiced, by the Israelites. Abraham evidently considered the phallus as an emblem of the divine, for he made his servant take a most solemn oath by laying his hand upon the master's genitals (under his thigh is the vailing translation). Jacob used the same form of obligation when Joseph promised to carry his father's bones out of Egypt. This form of obligation was used when "all the princes and the mighty men, and all the sons of David" swore allegiance to Solomon. These were important occasions when the most solemn obligations were taken; when levity or uncleanness could have no recognition; and when an appeal to the most sacred emblem of the divine would be made.

The same euphonistic translation occurs in speaking of the souls that came out of the *loins* of Jacob; and of the sons begotten of the *body* of Gideon; showing that it was common for the writer of the books of Moses to refer to the phallus as the source of children.

Abraham planted a "*grove*" in Beer Sheba, and there called upon the name of the Lord. Jacob set up a "pillar" and poured oil upon it, calling the place Beth-el — the house of God. Jacob also, in obedience to God's command to build an altar, set up a stone pillar and poured upon it oil and a drink offering of wine;

and again called the place Beth-el — the house of God. He also set a pillar upon the grave of his wife Rachel. When Jacob and his father-in-law, Laban, made a treaty of peace, they set up a pillar, and piled around it a heap of smaller stones; and while the former gave it one name and the latter another, still each in his own language called it the "Heap of Witness." Joshua, when about to die, took a great stone and set it up under an oak that was near the sanctuary of the Lord, saying: "Behold this stone shall be a witness unto us; for it hath heard all the words of Jehovah which he spake unto us." Samuel set up a "stone of help." All these things were done by men exemplary to the Jews; and the context shows that they are spoken of approvingly. Jehovah looked upon the Egyptians through a pillar of fire and terrified them; he led the Israelites by pillars of fire and cloud; he appeared to them in a pillar of cloud; came down in a pillar of cloud; Jacob calls him the shepherd, the "Stone of Israel;" Moses speaks of him as the "Rock of our Salvation" —the rock that begat thee — he is a rock. Samuel uses the same symbology. David says Jehovah is my rock. Elohim is my "rock" and "high tower" in whom I trust.

These allusions to Jehovah and Elohim, under the names of stone, rock, tower, high tower, pillar, etc., might be much extended; and, while they have all been interpreted in quite a different way, they are clearly phallic in their origin, as will more evidently appear when these symbols are spoken of as desecrated by being used in honoring other gods than Jehovah and

Elohim. The objection of the Jewish cult and prophets was not to the use or recognition of these symbols to represent the divine; but to their profanation in making them images or representatives of "strange gods." The objection was not to the symbol, but to the interpretation; for Isaiah says: "In that day shall there be an altar to Jehovah, in the midst of the land of Egypt, and a '*pillar*' at the border thereof, to Jehovah, and it shall be for a '*sign*' and a '*witness*' to Jehovah."

The command in Deuteronomy is not against planting groves nor setting up statues (pillars); but against such groves and pillars as "Jehovah hateth."

The worship of the sun and moon and of fire and water are always of phallic origin — and with phallic interpretation; hence the prohibition of this worship in the Mosaic law showed that it was a practice to be discontinued. Notwithstanding this law, we find that the kings of Judah built temples, ordained priests, and organized a system of sun and moon worship as gorgeous and sensual as that of the other Oriental nations, with all the accompaniments of horses, chariots, "groves," eunuchs, kedeshim and kedeshuth.

Moses was commanded to destroy the altars, break the pillars, and cut down the groves of the heathen tribes. Notwithstanding these plain commands, however, the children of Israel did evil serving Baalim and the groves; they also built them high places, and standing pillars and groves on every high hill and under every green tree; and they burned incense in these high places. The kings of Judah went so far as to

ordain priests, of whom there were four hundred and
fifty, for the burning of incense in the worship of Baal
in the courts of the temple and in these high places
dedicated to this idolatry.

The "groves," in the plural, were the lips of the
yoni. They were made of wood (sometimes of stone)
and carved as images. Gideon used this wood with
which to offer a burnt offering. They usually stood in
high places under green trees. One was in the temple.
They were sometimes surrounded by hangings or cur-
tains, forming tents, in which the worship of the groves
was participated in by both sexes, with the most licentious rites — under the direction of four hundred priests.
Solomon built "high places" for the worship of
Ashtoreth, Chemosh, and Molech.

The worship of Baal and Ashtoreth was not only
phallic, but sensually and broadly so — and, in some
cases, disgustingly revolting — and required the most
intimate and licentious association of the sexes. Baal-
Peor — which signifies God, the opener of the maiden's
hymen — was represented sometimes with a greatly exaggerated phallus, and sometimes with that organ in
his mouth. Philo says the devotee of Baal-Peor presented to the idol all the outward orifices of the body.
Another authority says that the worshiper not only
presented all these to the idol, but that the emana-
tions or excretions were also presented — tears from
the eyes, wax from the ears, pus from the nose,
saliva from the mouth, and urine and dejecta from
the lower openings. This was the god to which

the Jews joined themselves, and these, in all probability, were the ceremonies they practiced in his
worship; and added to their prostitution and dis-
gusting offerings their own children as a burnt
sacrifice.   One of David's mighty men was called
Baaliah or Bealiah — Baal is Jah; which would seem
to indicate that David was not strenuous in his opposition to Baal.   And David, on the most solemn occasion of bringing the Ark of God to the Holy City,
performed a most phallic ceremony of dancing in a
nearly naked condition, in the sight of the ark and of
all the people.   When his wife, Michal, sarcastically
chided him for this wanton display before the hand-
maids of his servants, he replied that he would "play"
and be yet more vile before them; and even that he
would be base in his own sight.

Samuel finds no fault with David for this phallic
procedure, but tacitly indorses it; for Michal, he tells us,
was cursed with barrenness — for her adverse criticism.

This illustrates again, as before said, that whatever
opposition there was to the symbolism and ceremonies
of worshiping Baal and other gods by the Hebrews,
still the great wrong, in the eyes of those who rebuked
or destroyed it, was not in this symbolism or in the
ceremonies generally, incident to that worship, for
many of them were common to the false worship and
to the worship of Jehovah or El.   The great impurity
consisted in the worship of these "strange gods," instead of bestowing all adoration upon the Hebrew god,
Jehovah.

The Jewish law says: "There shall be no whore (*kadesh* in the original) of the daughters of Israel, nor a sodomite (*kadeshuth*, masculine — and usually castrated) of the sons of Israel. Thou shalt not bring the hire of a whore (*zanah* in the Hebrew) or the price of a dog (*celeb*) into the house of the Lord." Here we have entirely different words in the same connection to mean those who practice promiscuous sexual union. The primary meaning of kadesh is "a consecrated one," and is used to indicate one who serves at or in the temple of worship; and it has both the feminine and masculine form indicated by varying terminations. This law does not prohibit this class, but declares they shall not be Israelites. These classes not only existed in Israel, but they were probably attached to the temples of worship by one set of authorities, who are blamed; and those who removed or destroyed them are commended for so doing. The women of this class wore a special attire, including a veil; and conducted themselves quietly — not seeking customers, but waiting for them to make the first approach. Tamar was thus arrayed when Judah thought she was "a consecrated one," or "temple attendant," or "religious harlot," and consequently one with whom he was legally permitted to associate in satisfaction of his passion; and the settlement of the matter indicated that he was excused, if indeed not wholly justified. The kadeshim and kadeshuth are supposed to have been the occupants of the small apartments attached to the temple or tabernacle, and were at the service of any one who desired and could pay for the accommodation ; and, as both sexes were included among them, their patrons could relieve the monotony of legalized fornication by the practice of tolerated sodomy. They occupied among the Jews at that time about the same place that "women of the idol" or nautch girls do among the Hindus. They were, no doubt, "the women who assembled in troops at the door of the tabernacle," with whom the sons of Eli openly and notoriously cohabited. The *zanah* — literally, semen emitter — was, on the contrary, an outcast, wearing a conspicuous attire, without a veil ; and was so bold of demeanor as to rush up and kiss men in public. The *celeb* — dog, sodomite — was a despised and execrated character, with whom no one acknowledged any relationship. These outcasts were, therefore, in wonderful contrast with the honorable attaches of the tabernacle — the kadeshim and kadeshuth.

All this does not, of course, indicate that the Mosaic law justifies or excuses these things. It simply illustrates that as a people the Jews were, in their lapses from rectitude, given to the worship of phallic gods, using phallic emblems, and engaging in phallic ceremonies — as licentious as other neighboring nations. Hosea, Jeremiah, Ezekiel, and other prophets are direct in their charges of these kinds of worship and licentious practices. Josiah found them all in full flower at Solomon's temple in Jerusalem, in Samaria, and in "every high place," and "beneath every green tree;" and his praises are sung for destroying the paraphernalia and idols, driving out the kadeshim and kadeshuth, and slaughtering the priests of this unholy worship. That is, he killed the provincial priests, but spared those in Jerusalem — probably because they were so popular that he dare not go so far in the metropolis.

When Rachel left her father's house, she carried away her father's terephim; David was in possession of such images; Micah made some for himself, which the Danites took from him, and which they worshiped as their god. These terephims were images of a man with phallus prominent and erect. Some of them were simply phalli, or the masculine triad. Maachah was deposed from being queen because she made a similitude of a phallus and worshiped it in a "grove." And Ezekiel charges this worship upon the Israelites.

Circumcision as a religious rite common to many ancient and modern civilizations is so clearly phallic as to need no comment.

GREEK AND ROMAN PHALLISM.

The Greek religion was essentially Indian and Egyptian in its mythology, dogmas and ceremonies. The Greeks, however, were not only extensive but very complimentary borrowers; for they gave to everything they copied from others a new lustre and an enhanced attraction by clothing it in new beauties.

The Greek myths, while essentially the same as those of the Hindus and Egyptians, and while, therefore, quite as phallic, were yet so logically constructed

and so poetically expressed that their superior con-
sistency and greater beauty made them seem more real,
and, therefore, more divine. Their worship was quite
as sexual as that of Phœnicia and Assyria; but it was
inculcated in language so impressively rhythmical, and
in allegories so hopeful and joyous of the heroic sac-
rifices and achievements of its deities, that it at the
same time charmed the ear with its melody and aroused
the imagination by its brilliant suggestions — while it
warmed the heart into grander enthusiasm and to di-
viner aspirations. Their ceremonies were as licentious
as those of Babylon and the Sactas; but they were
dressed in such attractive splendor, with a dramatic
movement so enticing and impressive, using a symbolism at once so realistically beautiful, and so pregnant of possible esoteric unfolding, conducted by a
priesthood grand in physique, cultured in intellect and
eloquence, and unsurpassed in graceful dignity, and,
in accordance with a ritual, so rich in the vitally and
actively beautiful, so well calculated to arouse enthusiastic and heroic ardor, and so full of charmingly
sentimental as well as subtilely amorous suggestions,
that the devotees — at least many of them — were so
exalted in their worship as to consider the sensual indulgences, and licentious rites in which they reveled,
as incidental adjuncts — rather than the fundamental
object of their Bacchanalian orgies.

It is scarcely possible for some minds to conceive it
possible that so much sublimity of real purity in
dogma, and so much of all that is beautiful in poetic

expression — both in word painting and statuary — was connected with such sensual ceremonies, and that priests and people alike engaged in such licentious and even unnatural sexual excesses.

"In Homeric days," says Mr. Gladstone, "we find among the Greeks no infanticide, no canabalism, no practice or mention of unnatural lusts; incest is profoundly abhorred. There is polygamy, but no domestic concubinage — and adultery is detested."

Among the sublime teachings of their grand philosophers, who are even now venerated for their transcendental utterances, and who had been initiated into the mysteries and helped to initiate others, and, hence, of course, participating in all the Eleusinian and Bacchic orgies, are the following : —

" The misery which a soul endures in the present life, when giving itself up to the dominion of the irrational part, is nothing more than the commencement, as it were, of that torment which it will experience hereafter — a torment the same in kind, though different in degree, as it will be much more dreadful, vehement, and extended. He who is superior to the domination of his irrational nature is an inhabitant of a place totally different from Hades." (How like St. Paul saying " our citizenship is in Heaven.") "They come to the blissful regions, and delightful green retreats, and happy abodes in the fortunate groves. A freer and purer sky here clothes the fields with a purple light; they recognize their own sun, their own stars."

Socrates says: " It is the business of philosophers to study to die, and be themselves dead; " and yet at the same time reprobates suicide; which is simply synonymous with Peter: " that we, having died unto sins, might live unto righteousness." Yet Socrates was a phallic-worshiping Greek; for, while he was not an initiate, as were his pupils, Plato and Aristides, he approved of the mysteries.

A great teacher has said: " The moral quality of human action does not lie in the particular thing done, nor in its effects upon the actor or upon others, but in the intention or motive of the one who acts." The great teacher of the Indias said, in relation to those not his avowed followers: " If they do it with a firm belief, in so doing they involuntarily serve me. I am he who partakes of all worship, and I am their reward." Greek instructors taught that ecstacy was sought as a state in which to receive divine influx; because, in this ecstatic condition the human soul pierces beyond the encumbrance of the body and enters into communion with the gods. Some of their writers tell us what they learned in this exalted and enthusiastic state: —

" I was taught that God is self-generated mind." " I saw that love was the first creation of the gods, and that from the divine influence of this impulse all that is created flows." " The great phalli at the door of the temple symbolize the divine activity which impregnates all nature."

Appuleius relates that during his initiation into the mysteries he " saw the sun at midnight." The literal reader disbelieves him or calls it a miracle. The initiate, however, does neither. He knows that Appuleius meant

that the sensual darkness of his natural mind was
lighted up to a perception of the higher truth while
looking upon the material symbols of the generating
deities.   Speaking of the sacred ark or cist of the mysteries, one said : —

"I saw in the egg the emblem of inert nature which
contains all that is, and all that is possible to be ; in the
serpent I beheld the suggestion of that divine impulse
to create which causes all productive action ; the phallus
glowed with supernal glory as I recognized in it the
exalted symbol of the creative gods, in generative
activity, producing the universe and all creatures that
are or will be."

It will be well to bear in mind these sublime ideas
and interpretations, and to remember the avowed intent of the mysteries and rites, while reading of the
gross procedures by which they sought to secure enlightenment and the favor of their recognized divinities ;
for surely the aspiring men of that day — like the same
class now — would often be led to feel — even if they
did not, like our later and more fortunate poet, sing or
say : —

> "But what am I?
> An infant crying in the night:
> An infant crying for a light:
> And with no language but a cry."

The Romans borrowed their religion largely from
the Greeks. That is, they borrowed the forms and
ceremonies. They, however, could not borrow the
poetry, sentiment, and enthusiasm. These are attainments which must be earned by generations of honest,

enthusiastic, and persistent study and practice. Such attainments are incompatible with a civilization — and impossible to the individual — like the Roman, in which the great ambition was military success, material aggrandizement and political preferment.

To give even an outline of Grecian and Roman mythology would require a volume, and, hence, only those classical dogmas and deities will be referred to which have a direct connection with the phallic ceremonies of their worshipers. Zeus is described as immortal and indestructible, male and female — androgynous. His head and face is the resplendent heaven, round which his golden locks of glittering stars are beautifully exalted in the air; on each side are two golden taurine horns — the risings and the settings — the tracks of the celestial gods; his eyes are the sun and the reflecting moon; his infallible mind is the royal and incorruptible ether.

Aphrodite, as the Celestial Virgin, and personification of procreative power, is represented — both in description and statuary — as a beautiful woman wearing a beard, and having at the same time a woman's breast — and sometimes locally double sexed. As the personification of amorous love or desire she is generally described and represented as a fully matured, young, and beautiful naked woman, of voluptuous form — and often, in posture and expression, or by holding a symbol, suggesting her passionate nature.

An image of Astarte was brought from Carthage to Rome, and there solemnly married to the emblem of the Sun-god. As these idols could not consummate the nuptials, the devotees, amidst rejoicing and revelry, acted as their proxies by engaging in a general and promiscuous orgy of feasting, drinking, and licentious indulgences. This, however, was only the European copy of the usual yearly Hindu celebration in honor of the mystic union of their male and female divinities.

In the temple of Venus, at Cyprus, that goddess was represented, in realistic detail, as androgynous; and her worship was there under the direction of castrated priests. Nor was this exception to excessive sexual indulgence an isolated case, for the priests of Dodona, the most ancient of the Greek oracles, were likewise eunuchs. The priests of the Orphic worshipers at Thrace were ascetics and devotees, and in many instances devoted virgins were required in the most sacred of their ceremonies and rites.

Jupiter, or Zeus, was represented crowned with olive, oak, or fir; his sacred color was white, and was worshiped in ceremony, partaking comparatively little of the phallic broadness which was bestowed upon his personalized representatives.

Bacchus — or Dionysus — represented the whole generative power. He was called "the father of the gods and of men," and "the begotten love." He was sometimes represented as androgynous, but usually as a male. He was called Choiropsale at Sicyon, Priapus at Lampsacus. Liber was the personalization of Bacchus as a mode of action — as Libera was of Venus. The goat was a special symbol of Bacchus; while

satyrs and fauns were his attendants or ministers.
Geese — and, hence, more poetically swans — were
sacred to Bacchus.

Priapus was represented as a man with an enormous
phallus; sometimes with a cock's comb and wattles.
He was also shown as Pan or a faun — with the goat's
horns and ears. When he had arms — which was not
always the case — the right hand held a scythe, and his
left often grasped his " divine symbol " — which was
always colossal, generally aroused and painted red.

Some of these Priapic figures, however, were not so
realistic and coarse. They were usually — if wood —
made from the fig tree, and often bore bells. Priapic
figures of the phallus or masculine triad, and these, in
association with the yoni, were common as amulets or
charms, and were worn either as jewelry in personal
adornment or in the bosom as charms to secure the
favor of the gods.

Greek and Latin authors make mention of the sacrifice of virginity to Priapus by means of a Priapic stone
or metallic phallus attached to an idol. And in some
places, at different times, brides, led there by their
parents, and in the presence of their newly married or
expectant husbands — take their first lessons in practical
Priapic worship, by means of the iron or stone symbol
of the sacred image, before being delivered to the husband's embrace.

There was found in Pompeii a bas-relief, in which
two elderly women — probably the mother and pros-
pective mother-in-law — were leading a young and nude

maiden to the " Hermes," by the phallus of which she
would give the gods the honor of her first experience
in coition. Generally, however, this ceremony was
simply the touching of the symbol with the *mons veneris* — or even pressing against it without raising the
skirts ; the actual initiation being in the orgies. Later,
however, the husband was supposed to be the real
initiator. This peculiar ceremony, like all the others,
was not a mere indecent procedure, but had a very
commendable object. The bride was thus brought to
the Priapic statue immediately before or after the marriage ceremony, and before its consummation, that she
might be rendered fruitful by this contact with the
divine generator, and be capable of faithfully and well
fulfilling all the new duties of her untried station as a
wife. An offering of flowers or a libation — generally
of wine — was often offered and special requests made
of the deity. It is reported that a lady — Lalage —
presented to the statue the pictures of Elephantis,
asking that she might be allowed to enjoy the passionate
pleasures over which he presided in all the positions
shown and described in that celebrated treatise ; and
the narrator remarks that, like a true devotee, she
probably strove to assist the god in securing a favorable
response to her prayers.

Married women also performed this ceremony in
order to destroy the spell that rendered them sterile ;
but — more experienced and less fearful — they carried
their devotions and the symbol farther — to actual introduction of the symbol into the vulva. Husbands

frequently accompanied their wives and saw that the ceremony was fully performed. A group in the gallery at Florence gives a representation of this ceremony. A woman, wearing a kind of cap, stands with her hands holding her uplifted garments. An enormous phallus rears itself from the ground and is shown in connection with her sexual organs — which are also exposed, and of unusually large proportions.

In relation to these Bacchic groups, as well as to Priapic statues, phallic amulets, and including the seemingly lascivious scenes upon vases, lamps, and other articles, it is clear that they were generally — almost without exception — religious objects, and hence not obscene in the sense of being designedly impure in their conception or use. They were used — as they have been found — nearly always in or about the temples — in or in connection with the tombs — or in the homes of the intelligent and the pious. Now, no people intentionally desecrate their tombs, nor of purpose aforethought defile their temples, much less would any people introduce recognized impurities among their children; and what is sacred in the sanctuary cannot be unclean or disgraceful in private life. In a word, these are religious emblems, and worshipful scenes. They were as common and as sacred among the Greeks and Romans as the cross and as scenes in the lives of the saints and martyrs are among Christians. The Priapic and yonic emblems were, as we know, symbols of divine creators and creation, and every composition into which they entered was interpreted from this key.

For instance, the phallus, bridled and ridden by a woman — her sexual organs also abnormally large, and exposed to view — is interpreted to symbolize Minerva bridling Pegassus, that is intuition — divine wisdom — the feminine side of intelligence, as guiding and controlling the creative energies and activities of the masculine generating powers and processes. Innumerable such instances might be cited, for the classics are full of them; and the reflective mind will easily find, what the poet and the mystic sees at once, the esoteric significance of every such symbol or group. Remembering this, and reading with this idea as an interpreting key, and the meaning of the group described in a former paragraph — in the light of the belief that such a ceremony would produce the desired result and secure a longed-for child — is readily understood. In the expression of that belief, and to secure that blessing, the ceremony is not only allowable, but commendable — sacred; and, hence, its representation is as pure as any other picture of a worshipful ceremony.

Considering the general state of reserve and restraint in which the Grecian women lived, it is to us of this day astonishing to what an excess of extravagance their religious enthusiasm was carried on certain occasions; especially on the celebration of the Bacchanalian orgies. The gravest matrons and the proudest princesses seemingly laid entirely aside their dignity and decency to vie with each other in revelry; they ran screaming through the woods and over the mountains, fantastically dressed or half naked, their disheveled hair interwoven with ivy or vine leaves and sometimes with living serpents. They frequently became so frantic as to eat raw flesh, and even to tear living animals to pieces, like beasts, with their teeth, and devour them while yet warm and palpitating. The religious rites of the Greeks, however, were generally calculated to arouse a joyous and festive enthusiasm. Their devotions were always accompanied with music and wine, as these tended to an exhilaration which assimilated the devotees to a like mind with the deity. They imitated the gods in feasting and drinking, in gladness and rejoicing, in cultivating and appreciating the elegant and useful arts, thereby aiming to impart and receive happiness.

The Greek women, singly or in groups, went to the temple or sacred places — that is, places made holy by the presence of a representation of a deity — and there made offerings to the divine emblem. This they did by wreathing the phallus with flowers, or anointing it with a specially prepared wine, or other compound, for the libation.

The mysteries of Bacchus were celebrated at Rome in the temple of that god, and in the sacred woods near the Tiber, styled Simila. At the outset women alone were admitted to those ceremonies — which were performed in the day time. Pacculla Minia, when made priestess, changed the nature and form of this worship by initiating her two sons and decreeing that the mysteries should be celebrated at night. Other men were introduced, and with them most licentious practices. The youths admitted were never more than twenty years of age. Wine, flowing in abundance, stimulated excesses, which the shades of night further favored.

The priests introduced the young initiates into subterranean vaults. Frightful yells and the din of drums and cymbals drowned the outcries which the brutalities inflicted upon the victims might call forth. Age, sex, and relationship were confounded. All shame was cast aside. Every species of luxury and sensual indulgence — even pederasty and Lesbianism — sullied the temple of the divinity.

If any of the young initiates resisted the importunities of the libertine priests and priestesses, or acquitted themselves negligently in the peculiar and often unnatural duties required of them, they were attached to machines which plunged them into lower caverns — where they met their death. Their disappearance was ascribed to the action of the angry deity whom they had offended by disobedience. Shouting and dancing, by men and women, supposed to be moved by divine influence, formed a leading characteristic of these ceremonies. Women with disordered hair plunged chemically prepared lighted torches into the waters of the Tiber without extinguishing them. At these midnight revels poisons were brewed, wills forged, perjuries planned, and murders arranged for. The initiates were of all classes — even the highest and most intelligent. Their numbers so increased that they were considered dangerous to the State, and the Senate abolished such assemblies.

General Furlong, in his "Rivers of Life," deals at great length upon the phallic basis of the religion of Rome. He says the Palatine Hill was from the earliest time dedicated to the male energy, while the Capitoline was especially sacred to the female cult — to which the Romans were, as a rule, the more favorable. The phallic emblems were afterwards modified or interpreted, so as to be adapted to the recognition and worship of Fire and Solar deities. Then, as now, women were the more enthusiastic and more active participants in religious devotions and ceremonies. St. Augustine (A. D. 400) tells us that the sexual member of man is consecrated in the temple of Liber, and that of woman in the sanctuaries of Libera — the same goddess as Venus — and that these two divinities are called the father and the mother, because they preside over generation.

Liber was a title of Bacchus, in whose honor the festival of the Liberales was held in March, six days after the Greeks celebrated their Dionysia, in honor of the same divinity. The phallus played a prominent part in these celebrations. It was, in some parts of Italy, placed upon a chariot, and with solemnity and great honor drawn about the fields, along the highways, and through the towns. At Lavinium the festival Liber lasted a month. During this time all gave themselves up to pleasure, licentiousness, and debauchery. Lascivious ditties and the freest speech were accompanied by like actions. A magnificent car, bearing an enormous phallus, was slowly drawn to the center of the forum, and there came to a halt. The most respectable matron of the town — as being worthy of this post of honor — advanced and crowned this symbol of the deity with a wreath. The more voluptuous part of the ceremony took place in the night — for it was considered unchaste to engage in this part of the rites in the day time. The worshipers gathered at the temple, where they lay promiscuously together, and honored their deity by a liberal display of the organs which represented him and his generative consort, as well as by their ardent and oft-repeated use in displaying the energetic and enduring powers which he conferred and blessed.

The next day, or at least soon after, each lady who had served as a sacrifice to the Priapic god by initiation into these experiences, expressed her gratitude for the benefits and pleasures she had received by offering small images of his characteristic emblem — equal in number to the men who had served her as priests in her sacrificial devotions. The number offered — as shown in some still extant medals, illustrative of this peculiar scene — indicates that the initiates were not neglected in this part of their devotions.

Some days later was celebrated the festival of Venus, also associated at Rome with the same emblem of virility. During this festival the Roman ladies proceeded in state to the Quirinal, where stood the temple of the phallus. They took possession of this sacred object and escorted it in procession to the temple of Venus Erycina, where they presented it to the goddess.

A Cornelian gem, with a representation of this cere-

mony upon it, was reproduced in the *Culte Secret des
Dames Romains.* A triumphal chariot bears an altar
upon which rests a colossal phallus. A genius hovers
above this symbol holding a crown of flowers sus-
pended over it. The chariot and genius are under a
square canopy, supported at the four corners by spears,
each in the hands of a semi-nude woman. The chariot
is drawn by bulls and goats, ridden by winged children,
and is preceded by a band of women blowing trumpets.
Further on — at the destination of the chariot and its
escort — is a symbolic yoni, corresponding in size with
the honored phallus. This female symbol is upheld
by two genii, who are pointing out to the approaching phallus the place it is to occupy.

When this ceremony was accomplished by the union
of these two emblems, the Roman ladies devoutly escorted the phallus back to its temple.

At the close of the festival of Venus came the Flor-
alia, which excelled all the others in license.

The prostitutes of the city mixed with the multitude
in perfect nakedness, exciting the passions by obscene
songs, jokes, stories, and gestures, until the festival
ended in a scene of mad revelry without the least restraint. Cato, the younger, who was noted for his
gravity, was present at one of these orgies, and there
was a hesitancy on the part of the participants about
giving reins to their inclinations; so out of respect to
the wishes of the representative citizens and matrons
he withdrew — so his presence need not interfere with
their worship or lessen their enjoyment.

A thousand sacred prostitutes were attached to the temple of Venus at Corinth, and a similar number to the temple of the same goddess at Eryx. Other temples in Greece were likewise furnished. St. Paul's description of the licentious practices at Corinth was, in a degree at least, true of most temples of Venus at that and some former times. Juvenal tells as that every temple in Rome was properly designated as a licensed brothel.

The Bona Dea seems to have been a more select society — a club, as it were, of the *elite* of Rome — organized and controlled by the *bon ton* of the Roman matrons. These Roman ladies were remarkable for their gravity, dignity, and virtue, in their ordinary life and associations. The stories told of them, however, relating their exploits of skill and endurance in the rites of Venus, show them to have been fully a match for the well instructed graduates of the seminaries of Corinth and Eryx; for they were experts in all the modes and attitudes which the luxuriant imaginations of experienced votaries have invented for the performance of the practical religious rites of their tutelar goddess. The ceremonies of the Bona Dea were a combination of all the rites of the other festivals. They were, however, as already suggested, participated in by the intelligent and prominent only, and hence were more elegant; and, while more refined in their procedures, were quite as free, licentious, and promiscuous — with all the revolting and unnatural practices of the more general orgies.

No very clear attempt has been made to unravel fully the Greek and Roman worship of Lares and Penates. They are in origin, however, strictly phallic. In India, at the present day, they are found in the niches of the domicile — elongated when they are Penates — in memory of male ancestors; and ovate when they are Lares — to commemorate the female dead of the family. The penates and lares — the phallus and the yoni conventionalized — commemorate the past vital fire and energy of the tribe or family.

NON-PHALLIC ZOROASTERISM.

The ancient Persians — under the teachings of Zoroaster — worshiped the good deity under the name of Ormazd. He was defined as goodness, intelligence, and light; and represented by the sun and the sacred fire. Ahriman — the embodiment of all evil, darkness and ignorance — was represented as night and winter. The feminine creator was represented by the moon, earth and water. The wind they recognized as the activities of these divine beings; good or bad — as it was beneficial or destructive, bringing pure air, comfort and health, or raging in storms and bringing destructive results.

In Zoroasterism we find the religion of the greatest purity of thought and ceremony among all the ancient cults.

In later times, probably 1000 or 1500 B. C., some of the Persians learned, and to some extent adopted, the Assyrian religion, and worshiped Mylitta under the name of Mithra — or the mediator — but never with gross licentiousness.

The followers of Zoroaster, the modern representatives being the Parsees of India, have never in their worship been gross or unclean in doctrine or ceremonial; and have never used any images of the Divine.

The serpent is spoken of as an evil principle, or as representing a servant of Ahriman, but never figured as a religious emblem by the followers of Zoroaster.

In short, the followers of Zoroaster were in no sense idolators. They were, from the beginning, as they are now, worshipers of one God. They held fire as sacred — not to reverence it for its own sake, but as the primal representative of the living and true, but invisible, God — creator of all that is.

Firdosi Toosi, the celebrated Mohammedan poet, who wrote *Shah Nama* — the history of the Persian Kings — placed on the title page the following verse as a motto:

> " Ma Pindar ke atush purustan boodund —
> Purustunduya Pack yezdan boodund."

> "Don't think they were fire worshipers;
> But worshipers of one God only." *

The Persian version of the fall of man is nearly like the Hebrew, but much more explicit: The first man, Meschia, and the first woman, Meschiane, were beguiled by the evil one, Ahriman, who appeared to them

---

* The author is indebted to Mr. SORABJEE ELCHIDANA, a learned Parsee, a native of Bombay, and now a resident of Los Angeles, California, for this translation, and for most that is here said concerning the religion of the followers of Zoroaster.

in the form of a serpent. Under his influence they committed the sin of carnal intercourse — in thought, word, and deed — and thus transmitted to all their descendants the taint of that sin. This myth, like one popular interpretation of the Hebrew legend, seems to many minds terribly inconsistent; for man was created male and female and directed to populate the earth — even to fill it — and was furnished with no means of doing so except the universal one of sexual congress. Yet these two interpretations make this union — the first obedience, to the first god, of the first command, to the first human beings — to be the first sin of those beings.

MIDDLE-AGE AND MODERN PHALLISM.

GNOSTICS. — Much has been said and written concerning the Gnostics. Some laud them as the wisest and purest — while others denounce and describe them as the most professedly and actually vile — among men. The simple fact is, that both these statements are comparatively true; because two entirely different schools assumed this name. One class were devoted students, austere, and abstinent, who mortified and reduced the body — crucifying the appetites and passions — in order to purify the impulses and elevate the mind. Of this class, while they were fanatical and unpleasant associates, much might be said that is favorable; but they do not come in the line of our work. Of the other class — who assumed the name in

13

self-sufficient arrogance — there were many sects. Their generally common dogma, however, was that there was no moral difference between human actions; and, hence, they made their religion minister to their greed and sensuality.

THE NICOLAITANS held that sensual pleasure was the true blessedness of man here, and the great end for which he was created; and that in the future life this realization would be immeasurably increased. Basilides was a fountain — or rather a sink — of all uncleanness. The followers of Carpocrates not only permitted sensuality and crime, but recommended them. Only those who daringly filled their measure of iniquity were saved; the only sin was in opposing the appetites and passions — which God had implanted; so their injunction was to yield to every carnal inclination, and their practices were in keeping with their doctrines.

One sect entertained the stranger with all the plenitude of bed and board; for, after the meal was disposed of, the host would arise and say to his spouse: "Go, exhibit to our guest your charity;" while he retired, that they might exercise their generating impulses.

Another sect revered and exalted Cain; and yet another held Judas Iscariot in the highest reverence. These Gnostics of the left-hand school cast the shadow of their errors and abominations over their purer and wiser namesakes; but, in reading of Gnostics, there need be no mistake as to which school is described.

THE NEZAIRES, or Nazarains, form — or at least very recently formed — a special sect in Syria. They adore God, and believe in Jesus as a prophet. They pray indifferently to the Apostles, the Virgin, and the ancient prophets. They practice baptism by immersion, celebrate the Nativity, the Ascension, and some other festivals — the most solemn of which they call the Festival of the Womb. In this solemnity they salute women with a holy respect, and affectionately embrace their knees, thus bringing the man's head on a level with the woman's abdomen. From this comes their title of Worshipers or Adorers of the Womb. They allow a plurality of wives and exalt libertinage into a moral maxim. On the day of the Circumcism — that is their New Year — all the women gather together in the hall of sacrifice. The windows are closed and the lights are put out. The men then enter, and each takes, by chance, the first woman he finds. This licentious ceremony is renewed several times a year, particularly at the Feast of the Womb. The chief and his wife at these times mingle on a level with the others.

ST. COSMO AND DAMIANA.— A very peculiar religious fair and festival was until quite recently held annually, on September 27th, at Isernia in Naples. The special feature of this occasion was that those diseased or weakened in any part of the body would offer a wax image of the part affected. This offering was accompanied by a fee to the priest and a prayer to the saint for restoration of health. Devout agents of the church mingled with the crowds, crying aloud " St.

Cosmo and Damiana," and carrying baskets full of these wax images. The price of these *ex votos*, as they were called, was "the more you pay, the more the merit." In the vestibule of the church were tables, at each of which presided a canon of the church crying, " Here masses and litanies are received," and taking the offerings presented. By far the larger proportion of " *ex votos* " are phalli or masculine triads of all sizes, and of varying shapes and conditions. Men — old, depleted, or diseased — offered counterfeits of the ailing or inert organs, asking for renewed health and vigor. The great majority of the devotees, however, were women and girls — widows, matrons, and maidens — who also presented *ex votos* of the masculine organs of generation — of all sizes, and in forms indicating health and vigor. These devotees paid the fee, offered their prayer, and, kissing the symbol, handed it to the priest. Among the prayers heard by an Englishman, who was at one time near a table, were the following: " St. Cosmo, dear Saint, bless me soon." "Let it be a boy." " St. Cosmo send him soon." " Dear Saint, let it be like this one," etc.

St. Foutin. — In some parts of France, until quite recently, St. Foutin received in some respects the same homage which was bestowed upon Priapus. This saint was credited with having the power of rendering barren women prolific, of restoring exhausted virility, and of curing venereal diseases. It was the custom of the men requiring his assistance to form *ex voto* in wax, representing the weak or diseased phallus. The women, on the other hand, made offerings of the phallus and its appendages in the form and of the size which they desired, in order to insure children.

Among the relics of the principal church at Embrun was the phallus of St. Foutin. The worshipers of this idol poured libations of wine upon its extremity — which was reddened by the practice. This wine was caught in a jar, allowed to turn sour. It was then called "holy vinegar," and was used by the women as a lotion with which to anoint the yoni. At Puy en Velay barren women prayed to this saint and scraped particles from the enormous phallus, of which they made a supposed fertilizing decoction.

At the church of St. Eutropius, at Orange, was an enormous phallus, and its natural appendages all covered with leather. This covering was removed when the barren devotees desired to worship it. At Bourg Dieu, near Bourges, the inhabitants worshiped a Priapic statue — probably of Roman origin. The monks, fearing the people, did not dare remove or destroy it, and so called it St. Guerlichon. Barren women flocked to this abbey, and, laying this statue upon the ground, stretched themselves at full length upon it. This was repeated for nine consecutive days. On each day they also scraped particles from the exaggerated phallus of this idol, which was soon very much reduced in size. The particles in an infusion was considered a certain means of overcoming barrenness. A similar statue stood in the chapel of St. Guignolé, near Brest. The very prominent wooden phallus of this saint traversed the statue, so that when the devotees reduced its size by scraping for their fertilizing decoction a mallet blow from behind performed the not seldom repeated miracle of restoring that important member to all its pristine size and glory. St. Gilles, in Brittany, St. Réné, in Anjou, St. Regnaud, and St. Arnaud were similarly adored. In the latter case a mystic apron covered the important symbol. This was raised in favor of sterile devotees, and a simple admiring inspection with proper faith was sufficient to secure the desired fertility. There are those who believe and suggest that the monks, as the living representatives of these virile saints, took an active and efficient part in rendering these devotions successful, by practically illustrating to these female devotees the method their husbands ought to follow in order to secure fertility. Whatever truth there may be in this suggestion, would only reflect upon the faithfulness of the monks, and not upon the Catholic faith.

Other cases might be cited, and, although this worship was opposed by the higher dignitaries of the church, they continued until the Revolution.

An enormous phallus of white marble, found at Aix, in Provence, was an *ex voto* offered to the deity presiding over the thermal waters by a grateful or expectant patient.

The bas-reliefs of the Pont du Gard and the amphitheater at Nimes show singular varieties of phalli — simple, double, and triple, with branches pecked by birds, furnished with claws, bells, etc. One is bridled, and ridden by a woman. A very singular and complicated monument of this worship was found in an ancient tomb near Amiens. It was a hooded human figure, in a walking attitude. It was in two parts. On removing the upper portion, consisting of the arms, head, and body, there remained an exaggerated phallus standing on the two human legs. This relic was preserved until the Revolution in the Chapter of the Cathedral at Amiens. In the museum at Portici, there is the cover of an ancient vase, which seems to have been used for religious purposes. On this cover is an enormous phallus, which a woman is embracing with her arms and legs. Another vase exhibits a dealer in phalli exhibiting his wares to a beautiful woman, who evinces evident delight at their extraordinary size and fine proportions.

THE JUDICIAL CONGRESS, sometimes spoken of, and oftener hinted at, as a practice of the seventeenth century in France, was a very simple affair. In those days, sexual excesses were common, while religious rule was rigorous; so that many poor creatures, with strong passions and keen consciences, were denied the gratification they so much desired.

If, therefore, one party in a marriage asked an indulgence or a separation, because of the impotency or inefficiency of the other party, the matter was sometimes referred to a select committee, who carefully examined into the matter. This committee were authorized to make occular and digital examination of the generative organs of one or both parties. They could, if they chose — and this they often did — order them to engage in sexual intercourse in the presence of the committee — this was called a "judicial congress" — so that the virility or impotency of either or both might be proven. The arbitrators had full power, too, to call in other parties — as experts or assistants — who would likewise in their presence engage one or the other of the disputants in sexual combat — in order to test their capacity and fitness for married life. Most of the complainants were women. The committee were, of course, men — and had the privilege of testing, as they deemed it necessary, the question of the woman's inordinate desire — or lack of proper desire — by a personal encounter with her. These trials, with their various modes, of "judicial congress," can, when these facts are known, be better imagined than described.

THE MAY-POLE. — The erection of the may-pole, surrounding it with wreaths of flowers, or gay and streaming ribbons, and dancing around it with merriment and roystering, sometimes ending in revelry and orgies, is a relic of the ancient custom of reverencing the symbol of creation, invigorated by the returning spring warmth. And it is realistically, as well as poetically, true, that

"In the spring a livelier iris changes on the burnished dove;
In the spring a young man's fancy lightly turns to thoughts of love."

THE LIBERTY CAP. — The mystic cap of liberty was originally red and a badge of citizenship, and, hence, of freedom from the many burdens and restrictions imposed upon foreigners. No foreigner was allowed to wear a cap of this shape or of a red color.

When an alien was adopted — or, according to American parlance, "naturalized" — he was circumcised, made a free man, and entitled to wear the "cap of liberty;" or, as it was then called, the "cap of circumcision." This cap, when cleft at the top so as to represent a fish's mouth, and, hence, the adoration of the Celestial Virgin mother, becomes the insignia of the royal priesthood, and is the official "*red hat*" of the Catholic cardinals.

St. Patrick and the Snakes. — When St. Patrick went to Ireland he found the people of that country much given to serpent worship, and their crosses adorned with that symbol — some of them very elaborately. He ordered these serpent emblems removed from the crosses. The clergy and most of the people obeyed the order. Out of this purification of the Catholic symbolism in Ireland arose the myth that St. Patrick banished all the snakes from the Emerald Isle.

The Fish is a well-known phallic emblem, symbolizing the feminine. It was used alone, and in many designs in combination with other elements, always, however, representing or referring to the Sovereign Goddess. Fish was, among many sects, an essential part of every feast in honor of the recognized deës, as well as the only animal food on days sacred to her service or worship.

Fish on Wednesday and Friday. — The eating of fish on Wednesday and Friday is usually explained as simply a sanitary measure — or as a fast for spiritual purification. This practice, however, was not originally a fast — but, on the contrary, a feast. It originated in the pagan practice of the worshipers of one cult of eating fish on Venus-day, or Wednesday, as we call it; while of worshipers of another cult, adoring the same goddess under the name Freya, had their feast of of fish on Freya-day, or Friday.

Among both these sects of worshipers it was a sacrilege to eat flesh on the " fish days " — or goddess-days.

The church adopted both these fish-eating feast days, giving both the days and the diet a very different value, and an interpretation more in harmony with its own doctrines.

THE MISTLETOE was dedicated to Mylitta, in whose worship every woman must once in her life submit to the sexual embrace of a stranger. When she concluded to perform this religious duty in honor of her acknowledged deity she repaired to the temple and placed herself under the mistletoe — thus offering herself to the first stranger that solicited her favors. The modern modification of this ceremony is found in the practice among some people of hanging the mistletoe, at certain seasons of the year, in the parlor or over the door, when the woman entering that door or found standing under the wreath must kiss the first man who approaches her and solicits the privilege.

THE DEVIL'S HORNS AND CLOVEN FOOT. — The idea and belief of the devil having horns and a cloven foot originated in the horns and cloven foot of the goat as a representation of Pan or Bacchus — the evil or false god — the devil — against whom the church had such an especially long and persistent fight. Nor was this identification of the Roman Bacchus and the orthodox devil an idle whim; for surely the practical and idolatious (adulterous) worship of Bacchus was, as it is still, one of the great evils which all true religions have most difficulty in overcoming.

DEVOTEES' NAMES.— When monks or nuns enter upon their consecrated life, they usually drop the names by which they are known to their associates, and assume a new name — as they enter upon a new life. This name usually also indicates the patron saint under whose especial protection they choose to labor and develop.

This is no new thing, for a prophet of old, when entering upon his mission, laid aside the name given him by his parents and adopted a new name, which, by its form and sound, indicated the god whom he served and whose truth he assumed to reveal.

Thus, Samuel, Ezekiel, and Daniel worshiped *El*. Jeremiah, Isaiah, and Hosea adored *Jah*. Joel acknowledged both *Jah* and *El*. Balaam adopted *Bel* and *Am*.

This is one of the keys to the prevalence of special cults, and often enables the truth-seeker to determine otherwise dark questions.

THE ORDER OF THE "GARTER"—the first of chivalry — is not a garter at all, but the "garder" or "keeper," the sacredest and secretest of woman's article of clothing. It is, by esoteric interpretation, unfolded to symbolize and emphasize the most exalted feminine virtue of chastity; and the one who worthily wins and wears this badge of knighthood should be the keeper and defender of the purity of every woman who needs his sympathy or his protection. It is the sistrum of Isis — the " cestus " or girdle of the immaculate virgin, the symbol of the divine woman which every man worships according to his idea of divinity and womanhood.